THE math EXPLORER

JEFFERSON HANE WEAVER

THE math EXPLORER

A Journey Through

the Beauty

of Mathematics

 Prometheus Books

59 John Glenn Drive
Amherst, New York 14228-2197

Published 2003 by Prometheus Books

Inquiries should be addressed to
Prometheus Books
59 John Glenn Drive
Amherst, New York 14228–2197
VOICE: 716–691–0133, ext. 207
FAX: 716–564–2711
WWW.PROMETHEUSBOOKS.COM

07 06 05 04 03   5 4 3 2 1

Library of Congress Cataloging-in-Publication Data

Weaver, Jefferson Hane.
    The math explorer : a journey through the beauty of mathematics / Jefferson Hane Weaver.
        286 p. cm.
    ISBN 1–59102–137–5 (pbk. : alk. paper)
    1. Mathematics—Popular works. I. Title.
QA93.W37 2003
510—dc21
                                                                2003012561

Printed in the United States of America on acid-free paper

This book is dedicated to my beautiful bride,
Shelley.

contents

preface

Why do we often find mathematics to be both fascinating and frustrating? For many of us, mathematics has an air of mystery with its sometimes forbidding nomenclature and its hieroglyphic symbols. But we also know from our everyday activities that mathematics is inextricably woven into our modern technical society. We use mathematics for a variety of tasks, such as buying and selling goods and services; building skyscrapers, bridges, and tunnels; tracking our financial accounts; constructing rockets, airplanes, and cars; and carrying out almost every kind of record-keeping activity. Indeed, it is more difficult to think of things that do not require the use of mathematics in some way than it is to designate those matters dependent on it. The most mundane measurements and the most complex calculations are both the province of the mathematician. But many of us are intimidated to some extent by this mathematical edifice even though we feel that we should be more adept at it.

This book was written to acquaint the reader with some of the more intriguing areas of mathematics, many of which may relate to our daily lives. It recognizes that mathematics has both an abstract as well as an eminently practical orientation. But this book is not a textbook and does not purport to examine all of the major topics of mathematics. Instead it invites the reader to go on a "walking tour" through mathematics and sample a few of the most interesting topics. The book begins with a discussion of both the origin and nature of mathematics. It then shifts to a discussion of both very large numbers and fractions. This is followed by a review of several of the most important areas of mathematics—algebra, geometry, and trigonometry—that existed before the invention of the calculus by Isaac Newton. Recognizing the fact that much of mathematics has to do with human behavior, the fourth part of the book looks at probability theory and introduces some of the basic concepts of statistics. Finally, the book shifts to a review of the life and work of the five men—Copernicus, Descartes, Kepler, Galileo, and Newton—who were most responsible for integrating mathematics with science and creating the foundations of our modern, quantitative physical science that continues to make possible extraordinary technological achievements. This seemed to me to be an appropriate way to conclude the book because what we now know as the scientific method, which has been responsible for so much of our higher civilization, would not have been possible without the achievements of these heroic individuals.

acknowledgements

I would like to thank Mark Christopher Louis Weaver, who created the graphics at the eleventh hour and made repeated changes with great patience and thoroughness. I would also like to acknowledge the many fruitful conversations I had with Adam von Romer on some of the subjects that ultimately appeared in this book. Finally, I would like to thank my editor, Linda Greenspan Regan, for her tireless efforts and valuable suggestions to improve this book.

part 1

THE WORLD OF MATHEMATICS

one

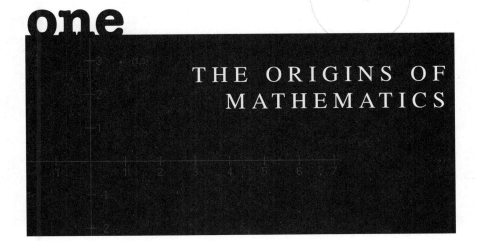

THE ORIGINS OF MATHEMATICS

Any curious individual will want to learn something about the origins of mathematics. For some people, such an inquiry is tantamount to a journey through an intellectual wonderland, replete with dazzling discoveries and wondrous breakthroughs. For others, a review of the beginnings of mathematics seems daunting. Between these two viewpoints, however, nearly everyone can expect to find his or her place and will undoubtedly benefit from learning something about the simplest operations of mathematics—counting, addition, and subtraction.

The first thing we need to realize is that mathematics as we know it did not suddenly appear out of thin air at a given point in history. Rather, it evolved in fits and spurts over time as various ancient civilizations in Africa and Asia fumbled and stumbled trying to formulate some of the most primordial concepts of modern mathematics. Unlike most of the physical sciences, however, the contributions of the earliest mathematicians have not necessarily been discarded in favor of more recent discoveries, but have often served as cornerstones for the creation of more sophisticated mathematics. More than any other discipline, mathematics is a cumulative process in which the newer mathematics must pay homage to its antecedents. This need to reconcile the new with the old also serves another purpose in requiring that all mathematics—old and new—has a certain basic consistency. If a groundbreaking new mathematical proof demonstrates that circles are actually rectangles, for example, then we would want to examine the most basic geometrical proofs to see if the celebrated Greek geometer Euclid or his colleagues had ever considered the geometric similarities of circles and rectangles. Because Euclid had compiled much of the knowledge of the ancient world about geometry in his *Elements*, we would probably be able to determine whether or not he ever made such a statement about the similarities of circles and rectangles. Indeed, Euclid took great pains to define each of the basic geometric concepts in detail. In doing so, he thus tried to make it possible for his readers to understand these concepts and their similarities and differences. Our review of Euclid's work would therefore compel us to reject this new mathematical proof and its ambitious objective of unifying two of the most basic geometric shapes because the structure of Euclidean (classic) geometry simply does not allow for such a melding of circles and rectangles.

COUNTING

Even though mathematics has a long and storied history, it took a long time before the ancients actually moved much beyond counting the number of fingers on their hands. There may be no written records of the first counting systems, but mathematics historians have surmised that the earliest counting may have begun with shepherds who needed some way to keep track of their flocks.[1] Some speculate that shepherds may have noticed that there was a one-to-one correspondence between each sheep and each member of a group of innocuous objects, such as the rocks or sticks in a pile on the ground.[2] In other words, they could match, item for item, the number of sheep and the number of rocks on the ground.

This is not so simple an operation as it may first appear because sheep do not usually stand still for very long, even when their shepherd is engaged in important mathematical work. So even though this shepherd had never heard of counting or the concept of matching objects in distinct groups in a one-to-one correspondence with each other, he could keep track of each member of his flock by moving one stone or stick for each animal that wandered past him. If all of the stones or sticks had been moved by the time the last sheep strolled into range, then the shepherd would know that all of his sheep had returned. Our lowly shepherd had no idea that he was discovering the most fundamental feature of mathematics—the countability of objects.[3]

This ability to match objects with each other—whether they are sheep, stones, sticks, animals skins, or spears—was profoundly important because of the universality of the process itself. One could match such things as the number of enemy warriors killed in battle to the number of original spears retrieved from the battlefield to the number of tents pitched in camp. The identities of the objects being matched, however, was irrelevant; the fact that each object from one group had a counterpart with an object from the opposing group (e.g., one enemy warrior could be match to one spear, the second enemy warrior could be matched to the second spear, and so on) was the real point and a central impetus for the development of mathematics and, indeed, organized knowledge as a whole.

It was only a matter of time before people decided that this matching process could be streamlined. After all, one might not have a flock of sheep on hand to match with the number of animal furs or tools or dogs one possessed. One observant person eventually decided that the ten fingers on his two hands were a useful group for matching with other types of objects. Unlike sheep, fingers did not wander off during the night. Flocks of sheep, by contrast, were far more unwieldy, required much time and effort to keep, and needed to be fed constantly. So fingers—almost by default—became the most common mechanism for matching objects in a one-to-one correspondence. The fact that most people possessed ten fingers is probably the reason we use a base 10 numeration system that utilizes ones, tens, hundreds, and so on. A base 2 system, by contrast, would use ones, twos, fours, eights, sixteens, and so on. As we have more than two fingers, however, there did not seem to be much point in considering a base 2 system. But the sheer drudgery of matching dissimilar groups item by item soon became evident. After all, it simply was not very exciting to match one's thumb to the first sheep, one's pointer finger to the second sheep, and so on.

This limitation of a finger-based counting system eventually prompted people to create an arbitrary set of numbers to represent each of the fingers on their hands. These numbers would eventually be described by using terms such as "one," "two," "three," and so on. But the leap would nevertheless be made from matching fingers with various groups of objects—which was of limited value—to the creation of an immutable set of conceptual or abstract numbers that could be used by anyone to "count" the number of objects in any group, whether it was sheep or oxen or any other potential source of animal protein. The advent of actual numbers offered a more abstract way to match the objects in any group with an arbitrary set of numerical symbols so that one could determine the "oneness" or "twoness" or "threeness" of a group. For example, one figure could be matched with one oxen, two fingers with two oxen, and so forth. In this way, fingers could be manipulated to represent classes of up to ten objects. Even though fingers themselves are not intrinsically similar to oxen, they nevertheless serve as a symbolic representation of the quantity of oxen being considered. It was this recognition that slowly led to a very rudimentary idea of the concept of number—separate and apart from the group of objects being considered.[4]

Now life would have been perfect if there were no groups of objects in nature having more than ten members because then a normal person could match his fingers with the objects of any group and feel as though he had mastered the world. But life is not so simple because the universe encompasses many objects, from grains of sand on beaches to the stars in the evening sky. Although one could expand the range of this counting system by adding toes, one would still be severely limited in one's ability to count any group containing large numbers of objects. Fortunately, someone hit upon the idea of using marks to represent objects. One could thus make as many marks in the sand as one wished in order to "count" the number of objects in a group; if one desired a more permanent record, one could cut notches in a stick or a piece of bone.[5] But the making of marks, particularly when one was dealing with very large numbers such as the number of bushels of corn produced in the kingdom or the number of people working in the fields of the monarch, was often exhausting and subject to error. At this point, some anonymous individual must have realized that one could dispense with the marks in the sand altogether by adding these marks together in his or her head. This may have been the first white-collar job where one could avoid physical labor under the pretense of needing to devote his or her energies to the demands of purely supervisory activities. Certainly this resort to the intellectual counting of objects must have seemed to be quite a timesaver for those persons who tired easily or suffered from severe hand cramps when making their marks in the sand. In effect, one could always create a larger number by adding one additional mark to the existing group. But the development of the first complete counting systems had to wait until each successive mark was assigned both a specific name (e.g., seven, eight, nine) and a specific ranking or position in the numeration system (on the number line).

The development of the modern counting system also gave birth to the concepts of ordinal and cardinal numbers. All cardinal numbers are numbers in unordered collections such as the number

of marbles on a table. You can count the number of marbles rolling
around the table but such counting does not enable you to draw any
meaningful information about the relative rankings of these marbles. Marble 7 is merely counted before marble 21. The counting
progression might have just as easily gone in another direction but
it would have made no real difference in the process itself because,
by definition, cardinal numbers have no inherent order. Ordinal
numbers, by contrast, are ranked in some type of order (e.g., fif-
teenth, sixteenth, seventeenth). If you were picking up several marbles from the table, then it would make sense to speak of the fifth
marble or the eighth marble you selected because there is an
inherent order in the selection process. The order of selection might
be important if we had agreed to pool our collections and that I
would select the first marble, you would select the second and third,
I would select the fourth, you would select the fifth, and so on. We
would therefore have agreed to an order of selection that would
necessarily require that we rank the desirability of the marbles to
us. For those who do not like to play marbles, this example may not
be completely satisfactory, but it is helpful in understanding the
idea of number as both a concept and a representation of rank in a
group. Both elements are essential for understanding numbers and
counting. Indeed, it was only after both concepts were taken into
consideration that any progress could be made in the development
of mathematics.

ADDITION, SUBTRACTION, AND THE CONCEPT OF NUMBER

Once these ancient mathematicians mastered the basic idea of
counting, then it was only a matter of time (centuries, perhaps)
before they began to group objects together and carry out the arithmetic operations that we now know as addition and subtraction.
Certainly basic arithmetic is an integral part of our early education.
For many of us, the mastery of addition and subtraction will mark

the culmination of our adventures in mathematics as we will now possess the basic skills needed to count our pennies and tally our income. But the actual operations of addition and subtraction, though fairly elementary in terms of their complexity, are part of the bedrock of mathematics and so it is worth a little of our time to explore them in some detail.

We can appreciate the idea that the act of counting may have been prompted at least in part by the practical limitations of using ten fingers to count large numbers of objects. This need was probably made even more apparent by the need to be able to combine or disaggregate quantities of objects. In other words, if one could count fifty items, then the need to be able to add another twenty items or take away forty items could hardly be avoided. Certainly merchants wishing to sell their wares to the general population needed to be able to keep track of their inventory in order to know when they needed to restock their stalls in the marketplace and also when to discontinue some of their more unpopular items. Of course it is impossible to provide an exact date as to when humanity began to use numbers in their businesses but it seems likely that it began before the advent of written records due to the fact that people engaged in trade and commerce in preliterate societies.

Being able to add and subtract items required people to develop some type of numeration system, which, as noted above, may have begun with shepherds tracking their herds. But as necessity is the mother of invention, we can surmise that the need to be able to tabulate the totals of objects—regardless of whether one was adding to or taking away from the collection—eventually forced the creation of a number system that could be used to add and subtract groups of objects from each other. This process of addition and subtraction can be represented in a graphical form by the use of a number line consisting of equally spaced numbered marks along its entire length. If one wished to add 4 to 7, for example, then one would start at the point marked 7 on the number line and then move to the right over 8, 9, and 10 before finally setting upon the 11 mark. Similarly, one would subtract one number such as 5 from another such as 8 by starting at the 8 mark and moving to the left five spaces past the 7, 6, 5, and 4 to the 3 mark. This use of the number line created a certain amount of excitement among mathematical thinkers because it provided a visible representation of addition and subtraction.

GREEK MATHEMATICS

The civilization of ancient Greece was the source of much of the mathematics and, in many cases, the sciences of our modern society. This is not to say that everything began with the Greeks because significant contributions were made by other civilizations such as the Egyptians and the Babylonians. But it is also true that the Greeks sifted through the knowledge in the ancient world, adopting some ideas and discarding others, until they were able to put a uniquely Greek imprint on contemporary mathematics. They were thus able to offer an incredible intellectual legacy that would affect not only the future development of mathematics but also the way in which the subject itself was treated by mathematicians and nonmathematicians alike. Perhaps as valuable as their actual contributions to mathematical knowledge, however, was the unprecedented degree of sophistication with which the Greeks approached the world and the topics of interest to them. The concepts of addition and subtraction offer a telling example of this striking divergence in perspectives that divides the Greeks from almost every other ancient civilization. Addition and subtraction represent little more than regurgitated mathematical facts for most individuals. In grade school, we are told $1 + 1 = 2$, $2 + 2 = 4$, and $3 + 3 = 6$. These facts are presented as absolute truths to be accepted without question. But most of us would be hard-pressed to recall any memorable explanation for why we should be learning these things in the first place—except that we would need to be able to add and subtract to be successful in our jobs and to carry out our daily lives. But to be fair to our teachers we can admit that most of us were not terribly interested as students in considering the underlying philosophical questions regarding addition and subtraction (e.g., Does the concept of number exist separate and apart from the object? Are addition and subtraction mirror-image operations of each other?). So it is perhaps a little surprising that the Greeks saw these most basic mathematical operations as having two aspects: the first, the *logistica*, referred to the arduous calculations used in everyday business and commerce, and the second, the *arithmetica*, can be defined as the philosophy of numbers.

Addition and subtraction, in their most basic forms, are merely mathematical operations whereby objects are collected together or taken apart and the resulting amount tallied by the operator. They are often taken for granted in the modern world. No doubt our intimate familiarity with counting operations has caused us to lose sight of the valuable role played by counting operations in our modern society. These operations are, fundamentally, a method for organizing things and grouping them together into discernible quantities. But many people would not be very impressed by this point. After all, what could be more obvious about the operations of addition and subtraction? But the Greeks, to their credit, saw something more important in these seemingly simplistic ideas. Aside from debating the merits of Plato's and Aristotle's views on politics and ethics, the Greeks engaged in lively banter about the concept of number and the relationships of numbers to the universe at large. This approach was very different from that offered by earlier civilizations such as the Babylonians and the Egyptians who were content to use basic arithmetic to count livestock and crops and the number of slaves needed to build hanging gardens and pyramids.

What was the catalyst that caused the Greeks to go beyond the rudimentary arithmetic of their predecessors and think about the nature of numbers themselves? Much of this credit is probably owed to Pythagoras of Samos (ca. 569 B.C.E.), a legendary figure who founded a school in Croton where he could pass on his peculiar brand of mathematical mysticism to his group of followers. Pythagoras and his followers were often forced to wander from place to place because Pythagoras had an uncanny ability to offend local sensibilities with his often controversial views regarding such things as communal living, women's rights, and the prospect of life after death. His was a society in which the members joined for life and were sworn to secrecy. But Pythagoras, for whatever reason, saw the universe as a reflection of mathematics as evidenced by his famous quote "All is number."[6] Although it is not clear whether Pythagoras had some cosmic-inspired vision in mind when he offered these words, Pythagoras was, in many ways, a symbol of a new approach to mathematics that focused on the theory of numbers and not merely the everyday use of numbers for making mundane calculations.

It is difficult to distinguish between the contributions of Pythagoras and those of other mathematicians in the ancient world who may have influenced him as he traveled to various places. Nevertheless, Pythagoras and his followers, perhaps because of their confidence that the study of mathematics and philosophy would provide a sort of code of conduct for life, considered numbers as being something more than purely mechanical devices. Whereas earlier civilizations had viewed numbers as nothing more than a means for solving specific problems, Pythagoras and his followers saw numbers as somehow transcending this mundane function to approximate something more akin to a search for knowledge and truth. Mathematics was thus both the means and the end, the all-encompassing enterprise that would provide them with insights into the human condition and beyond. But because most of the ancient texts about Pythagoras and his works have been lost, historians of mathematics are not quite sure which discoveries, subject to certain exceptions, should be attributed to Pythagoras and his followers. But the quantity of his individual contributions may not be as important as the depth of consideration which he gave to mathematics—particularly his own philosophy of numbers.

Pythagoras was only one among many of the early Greek thinkers who saw in nature a majestic, ordered, even mathematical design, a concept that was completely alien to the earlier Babylonians and Egyptians. Indeed, the Greek writers, such as Aristotle and Plato, acknowledged the contributions of the Egyptians and Babylonians to their own intellectual works. But these acknowledgments were overly charitable. In point of fact, the Greeks were the first to view the universe as rational even though their own mythology had allowed for a pantheon of gods whose flaws were altogether too clear. For those Greeks who were not so concerned with being struck down by one of Zeus's thunderbolts, however, the physical universe provided a tapestry from which they could draw their own mathematical designs and concepts.

But what was it about the Greeks that encouraged, and even compelled, some of them to discard the dark images of superstition and animism that had colored humanity's perceptions of the world? Simply put, it was the development of the process of logical reasoning and the elevation of ordered thought to the highest of priorities by the Greeks. This is not to say that the Greeks did not have time for amusements and other activities, as their respective city-states often engaged in bitter wars with each other with the victors enslaving the vanquished. While admittedly having a devastating impact on the Greek city-states, this continuing tendency to engage in ruinous warfare did not obscure the fact that the Greeks left an unsurpassed legacy of thought to the modern world.

The Greek approach to the study of mathematics, and indeed all knowledge, was one of enthusiastic curiosity coupled with a desire to understand the world in a way that was both coherent and unified. But the mathematics that we know today, though it originated in the ancient civilizations of the Middle East, is one that bears a uniquely Greek mold. As noted above, it was the Greeks who thought of mathematics as something more than simply a contrivance for adding and subtracting quantities. Instead they employed logic and reasoning to construct a mathematical edifice that was much more a product of analytical thought than simple trial and error. Progress in mathematics could now be attained by the use of intellectual inquiry instead of merely everyday experiences of grouping and splitting up collections of objects—whether they be ears of corn, articles of clothing, or sheep.

The Greeks also attempted to superimpose their mathematics upon the physical universe and seek to understand its mysteries through the use of numbers and geometrical shapes. Indeed, they saw the world as a manifestation of many different shapes and forms. But the Greeks diverged from their predecessors in choosing to focus on the abstract qualities of mathematics—that is, the perfect circle, the perfectly straight line—instead of the real-world manifestations of these shapes, such as the dome of a building or a well-traveled country road. They favored abstract and pristine concepts in their writings and accorded less favor to the actual physical counterparts of these conceptual forms and designs.

This preference for the ideal, as opposed to the real, fractured the philosophical connection between the Greeks and the earlier civilizations that had been content to build a rudimentary arithmetic based solely upon real-world experiences. Although this preference for the conceptual to the almost virtual exclusion of the real could be criticized, it is difficult to see how mathematics, could have otherwise advanced had such a fracture not taken place. Certainly the Greeks were mindful of the importance of the Egyptian and Babylonian mathematics but these civilizations did not provide a blueprint for moving beyond the most basic arithmetic operations because they were geared toward the most mundane of mathematical functions—the everyday operation of counting.

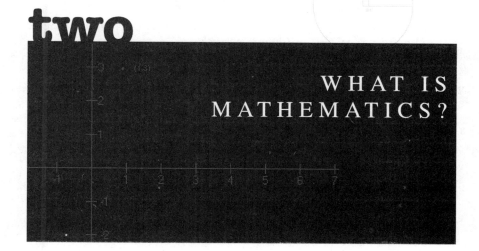

two

WHAT IS
MATHEMATICS?

Mathematics straddles both the familiar and the foreign. We use mathematics every day when we balance our checkbook (or leave it to the bank to figure out) or count our dollars out to pay a grocery bill. Many professionals such as engineers rely on higher mathematics like calculus to perform complex calculations and thereby determine such things as the strength of a dam or a bridge, the height to which a building can be constructed, or the amount of force per square inch that a pressurized tank can withstand before rupturing. Mathematics clearly has a variety of potential applications and, moreover, it occupies a unique preeminence in the pantheon of the sciences.

Professional scientists generally view pure mathematics as the supreme intellectual endeavor, ranking above the physical sciences such as physics and chemistry, because it transcends the physical world and provides the theoretical underpinning for modern science's attempts to describe and explain the universe. Mathematics provided the intellectual tools for formulating both Einstein's theory of relativity and quantum theory—arguably the two most important intellectual achievements of the past century. The ability to quantify scientific theories through mathematics has helped scientists to make verifiable predictions about the universe. Albert Einstein, for example, used mathematics to posit that space and time could be unified into a single four-dimensional continuum known as "space-time" and that a vast amount of energy could be released from the conversion of a comparatively small amount of matter—the famous $E = mc^2$ equation. In both cases, mathematics provided guidelines to physicists who were trying to understand certain fundamental characteristics of the universe. In addition, Einstein's general theory of relativity, which expanded his special theory to include gravitational forces and thereby showed that matter curves space, owed some of its mathematical pedigree to the work of the German mathematician Bernard Riemann. Indeed, Einstein himself was very charitable in the credit he gave to earlier mathematicians whose works would play such a pivotal role in his own theoretical research.

All of this sounds very noble but most readers who buy books about mathematics are not necessarily interested in formulating grand theories to serve as worthy companions to Einstein's theory of relativity. Yet many people would like to acquire at least some knowledge about the basic concepts of mathematics because mathematics and the language associated with it permeates our modern sophisticated society. But the quest for a greater understanding of mathematics does not entail purely practical considerations because many intellectually curious persons simply want to obtain a basic familiarity with some of the central ideas of mathematics. Perhaps this should be an objective of all persons who are interested in learning about mathematics and the ways in which it allows us to organize our perceptions of the physical world.

What is the point of trying to learn about the important concepts of mathematics? First, it provides you with certain insights into the dynamics of the physical universe because nearly all phenomena in nature can be described mathematically. Mathematics thus provides a quantifiable basis for predicting the behavior of natural phenomena. Second, a selective immersion in certain of the concepts of mathematics can help one get a more accurate "big picture" of the world in which mathematicians operate and thus enable individuals to understand the ways in which seemingly incongruous phenomena may be related to each other in subtle, often mysterious ways. At the same time, we can all begin to appreciate the ways in which mathematical concepts influence things that we would not associate with mathematics, such as art. Do we know, for example, why the paintings of the early medieval era look stiff and unnatural and the paintings of later eras look more natural and proportionate? The earlier artists did not know about the concept of the vanishing point whereby a painting could be designed using a series of lines all coming together at a single point on the canvas. By allowing for closer objects to appear larger and more distant objects to appear proportionately smaller, the artist could now recreate the scenes that he witnessed with his own eyes when looking down a city street or across a landscape in which the objects farther away appeared smaller to the eye. The vanishing point not only made it possible to paint with greater precision, it also offered a fine example of the role that mathematics—in this case, geometry—could play in the evolution of the arts.

DIGESTIBLE MATHEMATICS

Each of us has the capability of becoming more knowledgeable about mathematics if we are only willing to give up some time and roll up our sleeves. Now this may seem like a bold statement but it is true. Every person has the inherent ability to master many of the basic concepts of mathematics. The fact of the matter is that acquiring a certain fluency in mathematics does not require one to have an IQ of 200 or a Ph.D. or a complete command of the English language. But it does require the prospective mathematics student to have some respect for the subject matter and appreciate the need to think like a mathematician.

Now you may be wondering how one can think like a mathematician if the very mention of the word "mathematics" makes your palms feel clammy. But this does not pose an insurmountable barrier; it merely requires one to change one's general attitude toward mathematics. To think like a mathematician means that you view mathematics as an intellectual discipline as well as a very powerful tool that can be used to solve many kinds of problems that appear in almost every branch of human knowledge. You must also recognize that mathematical equations can be manipulated to yield an amazing variety of insights and information about our world. As a result, the mathematical way of thinking necessitates that you believe that you can employ mathematics to obtain solutions to almost any kind of problem that can be quantified. Unfortunately, many people who have been traumatized by high school algebra courses seem to view mathematics as a wild, uncontrollable beast or—alternatively—as an inscrutable subject having a language that can be understood by only a very few individuals. Such a view is unfortunate because it suggests that mathematics is inherently hostile and foreign. It would be far better to consider mathematics as a subject that all of us use to varying degrees in our everyday lives and merely acknowledge the fact that some persons are more adept in its applications than others. This apparent discrepancy in individual talents does not occur because mathematics itself is somehow hiding its secrets from all but a few "chosen" individuals, but instead because most persons simply stop trying to learn about mathematics once they have mastered the basic skills needed to carry out the mundane tasks of everyday life.

MATHEMATICS AND SUBJECTIVITY

Mathematics is an intellectual endeavor governed by precise, unchanging rules. It is therefore far more predictable and, in a sense, more comforting, than almost any other discipline. You will either have the correct or incorrect answer to a problem. There is no such thing as a mathematical answer that is "sort of" correct. This state of affairs is somewhat different among the liberal arts where subjective interpretations of a literary work, for example, can cause an answer or a response to be viewed by an all-knowing instructor as being anything from "marvelous" to "partially correct" to "possible" to "absolutely dense." Moreover, it is far more difficult to "politicize" mathematics; other disciplines, such as history, are far more susceptible to reinterpretation (though statistics can be selectively chosen based upon the personal biases and perceptions of the individual statistician). However, with history, any event can be viewed in a way that is colored by the observer's political and philosophical orientation. The Revolutionary War, for example, has been characterized in a number of different ways by people who have very divergent perspectives. Some viewed it as a gallant effort by the leaders of the American colonists to seek freedom from an oppressive Britain whereas other, less charitable observers have characterized it as a cynical attempt by these very same individuals to assert their own control over the new nation and perpetuate the dominance of the ruling class. Still others do not seem to know that we had a Revolutionary War at all.

Mathematics is the most abstract and conceptual subject ever conceived by the human mind. It is admittedly born of real world experiences such as the need to count items, but it is also governed by a set of self-consistent rules that impose order and predictability upon a seemingly chaotic world. But mathematics exists separate and apart from the real world and, as such, is both absolute and unyielding in its precision. One plus one is always equal to two in the mathematical world. It is not equal to three or five or seven as might be the case if something analogous to the revisionist historian existed in the mathematical world. It is this uniformity—the fact that the same equations will always yield the same answers time and time again—that arguably makes mathematics far more intellectually "cuddly" than its liberal arts counterparts. You always know where you stand in mathematics even though you may have not necessarily liked your position on the grading curve.

No student of mathematics will ever learn everything there is to know about all areas of mathematics. Indeed, no mathematician would ever make such a claim because mathematics has become so fragmented over the past few centuries that fewer and fewer mathematicians can claim to be "experts" in more than a few areas of mathematics. But even the greenest student of mathematics should recognize that he can educate himself about almost any area of mathematics, and, through a mastery of the rules, carry out almost any type of operation. The point of departure is the realization that the mastery of mathematics requires the would-be mathematician to learn a language that must be studied and understood just like any foreign language. It is this language that will make it possible for the mathematician to carry out particular mathematics operations.

Much of mainstream mathematics may be explained in a straightforward manner so that those who have not taken advanced courses in mathematics can appreciate the value and importance of these ideas. The calculus, for example, which is primarily concerned with quantifying rates of change and is absolutely critical for developing modern technologies, can be explained in a fairly simple manner using only a few symbols. A person who would like to learn about the calculus may not have taken courses in calculus while attending school but he can learn about the general principles of calculus by careful study of its basic procedures. Once mastering these basic ideas, he can bring his knowledge of the calculus to bear on a wide variety of technical equations that would have stymied many of history's great mathematicians who were simply not privy to its powers.

MATHEMATICAPHOBIA

What is it about mathematical symbols that causes people to freeze with fear? We have no problem solving the problem $4 + 5 = ?$ but many of us literally panic when we see a problem with alphabetic symbols such as $a + b = ?$. One would think that we would be equally comfortable with letters or numbers because many of us learn the alphabet before numbers. But there is something about using letters in a mathematical equation that causes many of us great discomfort and pain. The mathematical careers of many students run aground on the rocks of algebra because they are very uncomfortable with its formalism and its reliance on "foreign" letters for symbolic notation. Some may try to console themselves with the notion that algebra will never be of any real value in any career—except that of a high school teacher. But this conclusion would be dead wrong because algebra involves a form of abstract thinking that is a prerequisite for many professional and managerial occupations. For most careers that require any type of quantitative analysis (which also happen to be most of the better paying careers), you will need to have some familiarity with algebra and be comfortable with manipulating mathematical equations. And this trend can only be expected to continue with the increasing computerization of the workplace and the pervasive use of algebraic symbols throughout the corporate world. Many sales presentations, for example, will involve the use of charts with x (horizontal) and y (vertical) axes that will attempt to relate two or more changing quantities over time. These quantities could be any number of things such as calendar years, income, sales levels, and so forth. Anyone needing to give a presentation will have to be at least somewhat comfortable with discussing the relationships between the quantities described in the charts and be able to field questions about the factors that might affect these relationships.

Mathematics unquestionably has practical benefits. But mathematics is much more than merely a means to a successful career in sales or engineering. The successful student of mathematics should have some appreciation of the way in which mathematics is structured. And one can gain some important insights into this underlying intellectual exoskeleton by keeping in mind a few basic things: First, mathematicians prefer the simple to the complex. If one rule will do the work of two, then mathematicians will gladly take the one rule. Second, mathematicians seek rules of universal application. They are much happier when a concept such as the definition of a line as offered by the Greek geometer Euclid holds true in all circumstances; they become very skeptical when the definition of a concept must be continuously modified to fit different types of equations. Third, many mathematicians have an inherent bias in favor of immutable rules; they do not like change. This inherent conservatism has on certain occasions resulted in unwarranted delays in the widespread acceptance of important discoveries in mathematics. But it has also prevented mathematics from become overly polluted by "trendy" ideas that do not have a sound mathematical foundation. In fact, one can argue that most worthy ideas in mathematics—regardless of the skepticism with which they were originally greeted or the controversy that they aroused— eventually gain acceptance even though such acceptance may come two or three centuries after the deaths of the original proponents. Fourth, mathematics, though rooted in the experiences of the real world, is the most conceptual of all the sciences and its worth is totally dependent upon its internal rigor. Mathematicians use proofs to test the validity of particular mathematical concepts. This process subjects mathematical ideas to tests of rigor and consistency that simply do not exist in the arts. This testing process also reduces the spectrum of opinions in the mathematical universe. It is simply much more difficult to have widely divergent views about a particular mathematical concept that has been proven by competent mathematicians, such as the rules governing basic mathematical operations (e.g., addition, subtraction, multiplication, division). This stands in stark contrast to the world of literature in which subjectivity—not objectivity—is expected. This is not to say that sub-jectivity is somehow a lesser value than objectivity because there really is no such thing as pure objectivity in the world of art and lit-erature. Instead, people have opinions regarding the worth or worthlessness of a given painting, book, or movie, and their opin-ions will often range across the spectrum and range from utter dis-gust to profound adoration.

The point to bear in mind is that the arts necessarily arouse a wide variety of viewpoints and opinions, many of which may be diametrically opposed to each other but are still considered as per-fectly acceptable statements within the continuum among those involved with the discipline. The mathematician who fancies him-self as an iconoclast must work within the accepted structures of mathematics. If he is unable to provide the necessary quantitative support for his ideas, then he will be ignored or derided as a crackpot by his colleagues—regardless of whether his ideas are ultimately accepted as the greatest thing since the discovery of the compass. Indeed, it is fair to say that the idea offering only a mar-ginal or incremental improvement to the existing field of knowl-edge will stand a much better chance of immediate acceptance so long as it is solidly rooted in mathematics than the rare inspired idea that cannot be reconciled with existing mathematical para-digms and orthodoxy.

These points will assist you in becoming a more comfortable in your pursuit of mathematics because they give you certain clues about the ways in which mathematics evolves and cast light upon the contours of the mathematical landscape. After all, it is perfectly acceptable to have an earth-shaking idea so long as it fits within the boundaries of mathematics as we know it. Mathematics—to a cer-tain extent—is a social science in that mathematicians evaluate new ideas and may, regardless of the merit of these ideas, ignore them or even ridicule them merely because these concepts are revolu-tionary or unorthodox. The American philosopher Thomas Kuhn's *Structure of Scientific Revolutions* proposed that the evolution of science (and, by implication, mathematics) is governed by certain paradigms that determine to varying degrees the ideas accepted or rejected by scientists in a given era. But unlike the arts, the sciences—particularly mathematics—must eventually pay homage to certain quantitative relationships and, hence, be pulled back from the precipice of intellectual anarchy. This is really another way of saying that mathematicians generally share a view of their intellectual world that often compels them to try to fit new ideas into the existing framework like pegs in a pegboard. Some ideas arise as natural extensions of existing mathematical concepts and, if they meet the necessary standards of rigor and consistency, will probably be accepted as credible innovations. But those ideas that borrow from other disciplines or are otherwise not clearly of a classic mathematical pedigree may face tougher sledding.

It is my desire to take you on an exploration of many of the rudimentary areas of mathematics so you may share in the appreciation of the beauty and elegance of this encompassing discipline. What becomes apparent as one moves across the mathematical terrain, however, is that mathematics itself returns, in different ways, to the same underlying ideas. Similarly, the dependency of more elaborate mathematical concepts on seemingly simpler preceding ideas is a theme that recurs both throughout this book as well as the history of mathematics itself.

part 2

MATHEMATICS AND THE IMAGINATION

three

REALLY BIG NUMBERS

Many students of mathematics will, at some point or another, wonder whether they can count up to the biggest number and, indeed, whether they can even name the biggest number. This is not a silly question because most of us—even those of us who belonged to the our school mathematics clubs—are more typically used to dealing with quantities and amounts expressed in terms of hundreds, thousands, and even millions. But we do not often venture much beyond this range of numbers in our everyday experiences because we do not usually need to concern ourselves with the big numbers that invariably arise in trying to do such things as calculate the mass of the Milky Way galaxy or even count the number of grains of sand on the local beach. Certainly there are a few persons who do get to grapple with very big numbers (e.g., billions, trillions) on a daily basis such as the astronomers who estimate the number of stars in the observable universe or the government economists who calculate the gross national product, but these persons are more the exception than the rule. Most of us are more accustomed to dealing with smaller numbers.

The road to any big number is fairly straightforward in that it involves adding additional digits until the desired number is reached. In mathematics, we use very big numbers all the time, but there is a distinction to be drawn between those numbers that have been named and as a result adopted for general usage by mathematicians and those that are fanciful creations of anyone who wishes to string a bunch of numbers together for no particular purpose. For those persons who are interested in officially sanctioned numbers, we can refer to Table 1, which shows certain denominations of numbers beginning with one thousand and the number of zeros contained within that number.

TABLE 1: NUMBERS FROM ONE THOUSAND TO ONE DECILLION (UNITED STATES AND FRANCE)

Number		Zeros
Thousand	1,000	(3)
Million	1,000,000	(6)
Billion	1,000,000,000	(12)
Trillion	1,000,000,000,000	(18)
Quadrillion	1,000,000,000,000,000	(24)
Quintillion	1,000,000,000,000,000,000	(30)
Sextillion	1,000,000,000,000,000,000,000	(36)
Septillion	1,000,000,000,000,000,000,000,000	(42)
Octillion	1,000,000,000,000,000,000,000,000,000	(48)
Novillion	1,000,000,000,000,000,000,000,000,000,000	(54)
Decillion	1,000,000,000,000,000,000,000,000,000,000,000	(60)

The numbers in parentheses indicate the number of zeros that would be used for that particular number in Great Britain and Germany. The number one billion, for example, is equal to a thousand millions in the U.S. and French numeration systems and consists of a 1 followed by nine zeros. In the British and German numeration systems, however, the number one billion is equal to a million millions and consists of a 1 followed by twelve zeros. The basic difference is that each successive number (e.g., million to billion, billion to trillion, etc.) is defined in terms of a thousand of the preceding number in the U.S. and French systems but in terms of a million of the preceding number in the British and German systems as shown by the billion example noted above.

For most applications, the above table accommodates every need in solving mathematical problems relating to the known universe. But there are other bigger numbers that have been invented by mathematicians or, in some cases, the relatives of mathematicians. The name of one of the more whimsical numbers, the googol, was suggested by the young nephew of the mathematician Edward Kasner, and is written as a 1 followed by 100 zeros. But soon it became apparent that the googol concept could be taken one step further by offering a new number called the googolplex, which is a 1 followed by a googol of zeros. Now you may be wondering about the practical utility of having either the googol or the googolplex in one's mathematical vocabulary. Certainly one would be hard-pressed to determine how you might utilize the googol—let alone the googolplex—in the real world. Indeed, scientists estimate that the total number of elementary particles in the observable universe to be equal to a number that can written out as a 1 followed by about eighty zeros. As the googol itself is about 100 quintillion times as large as this quantity, the use of the googol here is not very helpful. If we wanted to use our existing nomenclature (e.g., billions, trillions, quadrillions, etc.), we could lump them together to get a number with eighty zeros. As a result we might express this number as a quadrillion decillion decillion. So we could say that the estimated number of particles in the known universe is about one quadrillion decillion decillion—give or take a dozen. But even though a quadrillion decillion decillion is a somewhat clumsy number, it is still nevertheless far more streamlined than writing out a sequence of eighty zeros. But we shall see that there is yet another, far more efficient way to express even the largest numbers with ease. Such a form of notation is absolutely essential to any mathematical student for whom time is a precious commodity.

Those of us who have only a passing familiarity with large numbers might find ourselves hard-pressed to determine the world's biggest number. So we would have to begin with a basic understanding of the way in which large numbers are created and used by mathematicians. Of course we are familiar with the most basic numbers such as ones, tens, hundreds, thousands, and so on. Once we get past millions and even billions, however, many of us may feel somewhat adrift because we have never had any reason to consider trillions or quadrillions or even greater numbers. But we need to bear in mind that it is not very difficult to create a larger number so long as we are willing to add enough zeros to an integer. Whether the newly created number is of any real value to anyone, however, will depend on whether it can actually be used for any meaningful purpose.

Meaningful purpose? A number? We need to bear in mind that the primary use of numbers in our everyday lives is to quantify particular groups or classes of objects. As a result, we can count the fifteen coins in our pocket, tally the $1,000.00 in our checking account, or take a headcount of the 10,000 people attending a professional tennis match. These numbers are significant because they are used to provide information about something in the surrounding physical environment. But if I was to unroll a sheet of paper five miles long on a country road and write a 1 followed by an average of ten zeros for every inch of paper, I would have written a number that would have a 1 followed by 3,168,000 zeros. This is certainly a very big number! Such a grand number would be worthy of a truly grand name such as—the "bazillion." Yet the reason that most people would not be leaping to their feet to sing my praises would be due to the complete uselessness of the bazillion to any one. After all, the world does not produce a bazillion apples each year, nor are there a bazillion fish in the sea, nor a bazillion birds in the sky. Indeed, the bazillion has no actual physical meaning at all aside from the fact that it can be written out on a five-mile-long strip of tissue paper. But it really cannot be used for anything at all such as describing the number of stars in the sky or the number of drops of water in the ocean.

Is it so essential, however, that there be something in the physical universe of that enormous scale in order for us to use exotic numbers such as the bazillion? It really depends on whether you believe that the number should have some application to a purely physical feature of the universe, such as the number of particles in the galaxy, or whether you are content with a number that has some arcane mathematical relevance, or whether you simply like the idea of playing around with very big numbers. From a purely mechanistic standpoint, one can create all sorts of big numbers simply by adding more zeros. A person could spend every waking moment of his entire life adding zeros to a number in a notebook in the hope of creating the largest single number known to the human race. But his masterwork could be undone in the blink of an eye by a single ne'er-do-well who grabbed the notebook and added a few more pages of zeros just for the fun of it. That, in a nutshell, is the problem that invariably arises in trying to create the biggest number—the fact that there are always bigger numbers lurking in the shadows that can be created with the addition of some more digits.

One thing to consider is whether any finite number—no matter how vast—should be consigned to the dustbin merely because it may not be used to describe some physical quantity such as the number of stars or atoms in our universe. In doing so, we are assuming that our own universe is ultimately bounded in some way and does not extend outward in every direction forever. Proponents of the "big bang" theory, which include most physicists, believe that the universe was born in a massive explosion twelve to fifteen billion years ago and that it will continue to expand outward until its gravitational field causes its expansion to stop and reverse. Then the system will collapse back to a superdense state (sixty to seventy billion years from now) from which the oscillation will begin anew. Because our knowledge of the observable universe is limited by the range of our largest telescopes, however, we are immediately forced to consider what lies beyond the limits of our own starry cosmos. Is the starry cosmos in which we live one of merely an infinity of galactic systems in a limitless dark void of empty space—with each such universe expanding and contracting over the course of billions of years and appearing as a mere speck of light in an unimaginably black ocean of nothingness? If our cosmic landscape is so vast, then it may be that our bigger numbers really do have some relevance because they are, ultimately, finite numbers, and as such should be useful for expressing any number of stars in what is, for all practical purposes, an infinite cosmic system. If we live in a trillion-galaxy universe, for example, and we assume that each galaxy has on average 100 billion stars, then we could suppose that the total number of stars in the known universe is equal to 100,000,000,000,000,000,000,000,000. But this is only the beginning because we then have to consider all of the other cosmos that could exist in this very same space for which—if we are dealing with an infinite space—can embrace any finite number we choose, whether it is the decillion, the bazillion, the googol, the googolplex, or all of them combined together. For as large as these numbers may be, they are still finite numbers. Consequently, they would all be contained within an infinite system—which is what we would have if our universe was but one amongst an infinity of other such universes. Although it is difficult to appreciate the difference between very large numbers having strings of zeros extending to the farthest known star and something that is truly infinite and therefore limitless, it is an important distinction that we need to try to keep in mind.[1]

EXPONENTS AND SCIENTIFIC NOTATION

Mathematicians routinely wrestle with large numbers and do not have the time or the energy or even the pads of paper available to write out every big number with all of their zeros. Indeed, it makes very little sense for them to bother with such a pointless exercise because there is a form of mathematical notation that allows anyone who cares to learn it to express very big numbers in a very concise manner. But before we get to this special form of notation—called scientific notation—we need to familiarize ourselves with the concept of the exponent itself. It is a concept that has made it much easier to manipulate unimaginably huge numbers with ease. To appreciate the usefulness of exponents to modern science and mathematics, we need to begin with the concept of the exponent itself.

The exponent has a peculiar feature when compared to other mathematical numbers we have considered because it always appears above and to the right of another number called the base. In the expression 7^2, the 7 is the base and the 2 is the exponent. The expressions tells us that the base 7 is to be multiplied by itself so that 7^2 is equal to 7×7, which is equal to 49. This simple example provides some hint as to the tremendous savings in time, paper, and ink that can be achieved with the use of exponents. After all, we were able to use this two-character expression 7^2 in place of the equation 7×7, which already puts us one symbol ahead in terms of savings.

It is indeed gratifying to report that exponents can be used to express quantities of any size and can certainly save the would-be writer of huge numbers from severe hand cramps. Suppose that we want to express the number 9,765,625 in the sleek exponent form because it is somewhat cumbersome to write it out completely. You could represent it with the number 5^{10} which would be a very good idea because 5^{10} is equal to $5 \times 5 \times 5 \times 5 \times 5 \times 5 \times 5 \times 5 \times 5 \times 5 =$ 9,765,625. Moreover, exponents can also be used in a variety of mathematical operations, such as multiplication and division. Suppose that I want to multiply $2 \times 2 \times 2 \times 2 \times 2 \times 2 \times 2 \times 2$ and $2 \times 2 \times 2 \times 2 \times 2$. One approach would be to retrieve a calculator and begin multiplying 2 by itself the indicated number of times. However, I could express this rather longwinded equation by the use of exponents and rephrase it as follows: $2^8 \times 2^5$. How do we solve this equation? We could multiply each expression by itself as indicated above and then multiply the two products together. We would find that our equation could be expressed as 256×32; the product of these two numbers is equal to 8,192. Or we could multiply the two exponents as follows: $2^8 \times 2^5 = 2^{8+5} = 2^{13} = 8,192$. Not surprisingly, mathematicians have created a general formula to express the multiplication of exponents that may be expressed algebraically as follows: $a^c \times a^d = a^{c+d}$.

Yes, we have just introduced an algebraic expression into the mix! Often such an expression will cause innocent bystanders to feel faint or to pass out entirely. The trick here is to stand your ground and look at the equation more closely to see that it is not really as scary as it first appears. First, we have to remember that this expression is merely an alphabetic version of the equation $2^8 \times 2^5 = 2^{8+5} = 2^{13} = 8,192$. The only difference is that we are using alphabetic symbols instead of numbers. What is the point of these letters? Well, it has very little to do with trying to intimidate algebra students and much more to do with describing certain relationships that will hold true for any integers (e.g., 1, 2, 3, 4) that are substituted in place of the variables a, c, or d. This should cause us to experience a sudden feeling of power and control over our physical universe because we can use this expression for any numbers we may desire. If you substitute many different numbers in place of the letters, you will see that this relationship holds true.

Of course no one would be content with merely multiplying exponents together because the entertainment value of such a procedure will diminish very quickly. So for a challenge, we can move to the division operation. As division is essentially the reverse of multiplication, so, too, is the division of exponents the reverse of the multiplication of exponents. As a result, we would not be surprised to see that there is another algebraic expression for the division of exponents that can be expressed as follows: $a^c / a^d = a^{c-d}$. As with our expression regarding the multiplication of exponents, this expression is extremely flexible in that it will work for any three integers that we decide to substitute in place of the alphabetic variables. The only difference is that the division operation necessitates that we subtract the second exponent from the first in order to obtain the answer. So if we go back to our original equation but alter it so that it involves division instead of multiplication, we will obtain the following result: $2^8 / 2^5 = 2^{8-5} = 2^3 = 8$.

Perhaps the most important point to be made is that anyone who is confronted with an equation consisting of letters instead of numbers needs to remember that numbers can be substituted in place of the letters so that the equation can be solved. Once you substitute the numbers in place of the letters, the equation will be solved and the shroud of mystery will disappear.

Now we know that the world of mathematics may not always be so clean and tidy. Indeed, we should not be surprised if we were to encounter a more sinister-looking expression in which we were asked to divide a smaller exponent by a larger exponent. Suppose that we have the equation $2^8 / 2^5 = ?$ From our previous discussion we know that we have to subtract the second exponent from the first. But this leads us to an uncomfortable conclusion: $2^5 / 2^8 = 2^{5-8} = 1 / 2^3$. More generally, we can state this expression in the following way. If c and d are whole numbers and d is greater than c, then $a^c / a^d = 1/ a^{c-d}$.

The more curious among us might wonder what would happen if we divided an exponent by itself. Let us consider the following expression: $2^5 / 2^5 = ?$ We would solve it as follows: $2^5 / 2^5 = 2^{5-5} = 2^0$. Hmm. What in the world are we supposed to do with our base 2 raised to the 0th power? Fortunately, mathematicians long ago anticipated this problem when they proposed than any number raised to the 0th exponent would have a value of 1. So we can complete this expression as follows: $2^5 / 2^5 = 2^{5-5} = 2^0 = 1$. There is another algebraic theorem that can be used for this expression: If c is any positive whole number, then $a^c / a^c = 1/ a^{c-c} = a^0 = 1$.

Having delved into the subject of exponents enough to satiate even the most ardent mathematics student, we now can bring our discussions of very large numbers and exponents together to reveal the way in which mathematicians can express extraordinarily vast numbers in extremely concise forms. This technique, known as scientific notation, is a wonderful timesaver for scientists, mathematicians, and others who must deal with extremely large numbers. Quite simply, it involves expressing any given number in terms of a number multiplied by a base (typically 10) raised to some exponent. For example, we could express the number 5,392,400 in scientific notation form as 5.3924×10^6. Because we are using a base 10 system, the exponent 6 tells us that the decimal point needs to be moved six places to the right in order to express the number in its conventional form: 5,392,400. To represent the number in scientific notation form, we simply move the decimal from the right of the last zero (in the ones place) to the left six places so that it is to the right of the first significant digit (in the millions place). As a result, we find we now have 5.3924×10^6.

We can even use this technique to express much bigger numbers with great ease. One quadrillion, for example, as shown above, can be written in its conventional form as 1,000,000,000,000,000. But if we simply do not have the energy or the patience to draw so many zeros, we can use scientific notation and express it as 1.0×10^{15}. Again, the exponent 15 tells us how many places we would have to move the decimal point to the right in order to convert this number back to its conventional form.

But this is only the beginning because we can sense that scientific notation offers unlimited potential for making short work of very large numbers. Without even breathing hard, we can leap forward and write 1 octillion as 1 followed by 27 zeros (1,000,000, 000,000,000,000,000,000,000) or we can save a great deal of time and effort by expressing it in scientific notation form as 1.0×10^{27}. The choice of using scientific notation or conventional notation when expressing large numbers such as an octillion or even a decillion or a googol is rather like deciding whether to take an attache case or a steamer trunk on the train to work.

Can scientific notation be used to express very small numbers? Of course. Scientific notation goes both ways and can also be used to express very small numbers such as 0.00000007 or 0.000000000 0000000000000456 or even 0.00000000000000000000000000000000 000000000000000000000012. Now you must be thinking that this scientific notation is wonderfully flexible to be useful for both the very large and the very small and you would be correct. As we now know, the exponent indicates how many places the decimal point should be shifted to the right when we are dealing with numbers greater than 1. When we are dealing with numbers less than one, the exponent is used to show us how many places the decimal point should be shifted to the left to express the number in its decimal form. The only difference as to whether we move the decimal to the right or the left is whether the exponent is positive or negative. So we can use scientific notation to express 0.00000007 as 7.0×10^{-8} or 0.00000000000000000000456 as 4.56×10^{-23} or even 0.0000 00000000000000000000000000000000000000000000000012 as 1.2×10^{-52}. Whether you actually have any real purpose in writing out such a lengthy sequence of zeros is up to you. But these examples illustrate the fact that scientific notation is a wonderful form of mathematical shorthand for expressing both the very large and the very small.

We mentioned earlier some of the very large finite numbers that mathematicians have utilized, highlighting their generally accepted names and magnitudes. But they do have their counterparts in the world of the very small. Table 2 illustrates the generally accepted names and magnitudes of the very small numbers:

Table 2: Numbers from One-Thousandth to One-Decillionth (United States and France)

Number		Scientific Notation
Thousandth	0.001	$10^{-3} = 1/10^3$
Millionth	0.000,001	$10^{-6} = 1/10^6$
Billionth	0.000,000,001	$10^{-9} = 1/10^9$
Trillionth	0.000,000,000,001	$10^{-12} = 1/10^{12}$
Quadrillion	0.000,000,000,000,001	$10^{-15} = 1/10^{15}$
Quintillionth	0.000,000,000,000,000,001	$10^{-18} = 1/10^{18}$
Sextillionth	0.000,000,000,000,000,000,001	$10^{-21} = 1/10^{21}$
Septillionth	0.000,000,000,000,000,000,000,001	$10^{-24} = 1/10^{24}$
Octillionth	0.000,000,000,000,000,000,000,000,001	$10^{-27} = 1/10^{27}$
Novillionth	0.000,000,000,000,000,000,000,000,000,001	$10^{-30} = 1/10^{30}$
Decillionth	0.000,000,000,000,000,000,000,000,000,000,001	$10^{-33} = 1/10^{33}$

As with the previous table, the orders of magnitude decrease by millionths instead of thousandths in the British and German systems. The numbers in the right column indicate both the conventional exponent representation and the fractional equivalent of these numbers.

As with the very large numbers, one can go on as long as one pleases and string together an excruciatingly long sequence of zeros to the right of the decimal point before succumbing to the temptation to write down some nonzero numbers and finish the job. If we return to the wonderful number we created earlier in the chapter—the bazillion—we could certainly create its decimal counterpart—the bazillionth.

How do we describe the bazillionth mathematically? We already know that the bazillion itself boasts a 1 followed by 3,168,000 zeros. Despite the reluctance of scientists and mathematicians alike to embrace the bazillion, we know that it can be expressed in scientific notation as 1.0×10^{3168000}, which is far easier than filling a warehouse full of notebooks with zeros. The bazillionth, not surprisingly, can be described in the same way but with a negative exponent: 1.0×10^{-3168000} or, expressed as a fraction, $\frac{1}{10}^{3168000}$.

Scientific notation provides a wonderful tool for anyone who must work with very large or very small quantities. It shows us those numbers that are significant for solving the problem at hand by disposing of the often numerous zeros that follow those numbers. If we are dealing with 6,330,000, then we really only have three significant numbers—6.33×10^6 because the numbers that follow these three in conventional form are merely zeros that identify the magnitude (millions) of our number. In essence, the recognition that we must work with significant numbers makes it possible for us to cast aside the zeros in favor of the most efficient exponential notation.

Exponents are very helpful in creating very big and very small numbers. If the term 1.0×10^{80} describes the number of elementary particles in our universe, then the term 1.0×10^{146} would describe the number of elementary particles in a decillion decillion universes similar to our own universe. This number may give us some insight into how seemingly innocuous-looking numbers expressed in scientific notation can be mind-numbingly vast.

But we can take exponential notation in another fascinating direction and create numbers that outstrip our ability to imagine them—even though these numbers are still ultimately finite in nature. Cambridge mathematician S. Skewes, in working on a problem of the theory of numbers known as the Riemann Hypothesis, offered a number that now bears his name in considering the assumption that the prime numbers become less and less common as we consider bigger and bigger numbers. At any rate, his number, known as Skewes's number, is written as

$$10^{10^{10^{34}}}$$

which, despite its rather harmless appearance, is actually quite ferocious in its ability to gobble up billions of notepads belonging to the mathematical students who would dare try to express it in decimal notation. Why is this so? We must remember that 10^{34} is equal to 10,000,000,000,000,000,000,000,000,000,000,000, which the mathematician Andrew Hodges in his thoughtful book *Alan Turing: The Enigma*[2] has estimated is equal to the number of elementary particles in a large building. But 10^{34} is "small potatoes" if one decides to consider $10^{10^{34}}$ which would be equal to 1 followed by 10^{34} zeros; in Hodges's view, this would "require books with the mass of Jupiter to print it in decimal notation."[3] But here again $10^{10^{34}}$ becomes very much of a "junior player" when compared to Skewes's number, which is incomprehensibly bigger and expressed as 1 followed by $10^{10^{34}}$ zeros.

Since you are alert, you are no doubt thinking that Skewes's number is not the be-all and end-all of the number system. And you would be right because you are a perceptive individual. Because you have undoubtedly noticed that one can always create a bigger number by adding an additional digit to any number, you may have also concluded that one could very easily surpass Skewes's number by adding a fourth level of exponent such as

$$10^{10^{10^{10^{34}}}}$$

which would be equal to 1 followed by Skewes's number of zeros. Not surprisingly, there is no rule that says you cannot continue as long as you like, adding level after level of exponents—hundreds, thousands, millions, billions, trillions, quadrillions, and so on—as long as you like. But the drawback to this game is that you quickly lose any comprehension of the magnitude of the number you are creating. I can certainly describe a number in which 10 is raised to the trillionth power, which we would express as $10^{1,000,000,000,000}$. I could then imagine it as being raised in turn to a second level of exponent to the trillionth power, which could then be raised to a third level, and a fourth level, and on and on until we decide to stop, having created a number in which 10 had been raised to the trillionth power 1 trillion times. Needless to say, Skewes's might be disappointed that our trillionth-level exponent had squashed his vaunted number so effortlessly. But I would have great difficulty in trying to understand this trillionth level exponent. Certainly the unending nature of this process was recognized by Hodges when he noted, in commenting on Skewes' number, the following: "In actual fact mathematicians had certainly thought about numbers far larger than these; here we have only gone through three stages of growth, but it is not difficult to make up a new notation to express the idea of going through ten such stages or 10^{10}, or $10^{10^{10}}$; or of regarding even these as just the first step in a process of super-growth, and then defining super-super-growth, and then. . . ."[4]

Every mathematics student can create his own number and can make it as large as he may desire but he must understand that the simple act of stringing zeros together will not create a meaningful number. Mathematics, for all of its theoretical underpinnings, is ultimately a discipline that measures and otherwise quantifies phenomena in the physical universe. Any supernumber that we create, regardless of whether it has seven, twenty, two hundred, or five billion levels of exponents, will, in all likelihood, be too vast to be of any real use to us. Of course there will always be a certain satisfaction that comes with creating a monstrously huge number. But the feelings of elation may be short-lived—much like the person who is the only one to show up at his own birthday party—because in mathematics one must actually be able to do something with that number. Otherwise, we may have nothing more than a numerical curiosity that has no more flesh on it than a ghost.

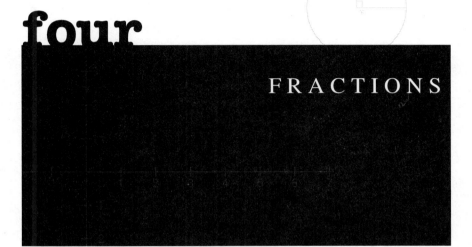

four

FRACTIONS

Because mathematics originally developed as a system for counting whole objects—bushels of wheat, rocks, trees— it always had a general orientation toward quantifying numbers of things. But life is not so simple that we can always count on having whole objects to consider when doing mathematical calculations. After all, we can scoop out handfuls of wheat from a bushel or break rocks into pieces or saw trees into logs. But we can scarcely talk about "parts of whole objects" if we do not try to attach some kind of mathematical description to these constituent parts. Indeed, it makes no sense to talk about "some" of the wheat or "a few" of the rocks from a mathematical standpoint. You cannot add "some" wheat to "more" wheat and be able to draw any meaningful conclusion. It is the need to be able to deal with the parts of whole objects that gave rise to the concept of fractions. Unfortunately, the subject of fractions presents a substantial stumbling block for some seeking to learn about the mysteries of mathematics. Even though the benefits of being able to perform mathematical calculations on parts of objects as well as the objects themselves would appear to be useful, some individuals find themselves unable to embrace the concept of fractions and accept the notion that fractions can be valuable for resolving everyday problems.

Why should fractions present an intellectual challenge to us? After all, we deal with fractions every time we have to leave change on a table at a restaurant because we usually base the amount of the gratuity on the amount of the bill itself. Coins can be clunky to handle when compared to crisp new bills and are cumbersome to carry. But coins, despite their bulkiness and weight, provide the perfect illustration of how fractions work. A one-dollar bill may be thought as having the value of one whole. A quarter is equal to 25 cents or one-fourth of the dollar. We know that four quarters—each of which are equal to one-fourth of a dollar—will total one dollar or one whole. A similar lesson can be drawn for pennies, nickels, and dimes. Each penny, for example, is equal to 1 cent or one one-hundredth of a dollar (0.01 in decimal form). Thus all of these coins represent fractional shares of the whole number one, and as such should be viewed as a working model of fractions by those who have an aversion toward numbers in general and fractions in particular.

Which ancient civilization had enough time on its hands to come up with the concept of fractions? Many mathematics historians generally attribute the concept of fractions to the Egyptians and the Babylonians.[1] But the verdict is not conclusive as to which civilization made which mathematical breakthroughs at a given time in history. Indeed, some societies showed confusion in their use of mathematical symbols, which suggests that their mathematicians were not always sure about the proper uses for fractions or how such symbols should be manipulated. According to Morris Kline's wonderful book *Mathematics in Western Culture*, "the Babylonians lacked adequate notation [to use fractions] and the correct value had to be understood from the context."[2] Therefore, the value of a particular mathematical symbol was not necessarily immutable but could vary depending upon the way in which it was used. Some persons, particularly those who believe that mathematics should be more democratic and allow for more than one right answer to a given equation, might applaud the versatility of the Babylonian nomenclature. Others, however, might find its inherent mushiness to be aggravating. The Egyptians were certainly more skillful in their use of fractions but, as Kline points out, "the Egyptians found it necessary to reduce a fraction to a sum of fractions in each of which the numerator was unity [so that] they would express ⅗ as ½ + ⅛ before computing with it."[3] Needless to say, this process of splitting up fractions into constituent fractions did little to promote the ease of use of fractions. And while this use of constituent fractions may not rank as the crowning glory of human civilization, it did represent a big step forward in the ability of humans to think of numbers as concepts and not purely as nametags for quantities of objects that they observed in the real world. Indeed, the concept of fractions is perhaps the first topic encountered by some students of mathematics that seems to be "unreal" or not have an immediately apparent "practical" application to everyday life.

But what is it about fractions that seems daunting to some persons? Perhaps it is the fact that we are not very intuitive about fractions. We can certainly understand that $1 + 1 = 2$ because this type of operation can be represented by successive movements along a number line (from 0 to 1 to 2) or a handful of fingers. But it may seem odd intuitively to divide a smaller number by a larger number. We cannot divine the answer simply by hopping along the number line from the smaller number to the larger number or vice versa because we would only be performing the addition or subtraction of two numbers. In other words, the move from 2 to 5 would be the number line equivalent of $2 + 3 = 5$, whereas the movement from 5 to 2 would be the number line equivalent of $5 - 3 = 2$. Moreover, we usually think in terms of "wholes" and not "parts" when we view the physical world. And this preference for whole objects obviously colors the way in which we view fractions and their usefulness in everyday life.

Perhaps fractions become less intimidating if we remember that they are merely a "special" kind of division in which the numbers are expressed in terms of one number being divided by another number. So ½, for example, is the result of dividing 1 by 2. Similarly, ¼ is the number that results from dividing 1 by 4. The important point is that fractions are the result of nothing more sinister than division.

Our past introduction to fractions typically begins with an explanation of the structure of fractions. In the fraction ½, for example, the 1 is called the "numerator" and the 2 is called the "denominator." The numerator is always on top and the denominator is always on the bottom. Our nomenclature simply refers to the fact that somebody decided to use those terms to refer to the numbers above and below the line of any fraction. A fraction is essentially a representation of the process of division whereby the numerator is divided by the denominator. In our fraction ½, the 1 is being divided by the 2. Some may initially believe that since 2 is larger than 1, 1 cannot be divided by 2. But it can. Any number can be divided by any other number, but the answer may not be a whole number like 1, 2, 3, and so on. With our fraction ½, we can divide the 1 by the 2, but we cannot get an answer that is a whole number because 2 simply does not divide equally into 1. Upon further consideration, we find that 1 divided by 2 equals ½ or, in decimal terms, 0.5. Why? Well, 1 (the smaller number) is being divided by 2 (the larger number). As a result, the answer will necessarily have to be less than 1. How much less will depend on the two numbers being considered. In the case of this fraction, we will find that 1 divided by 2 is equal to one-half or, as shown by the fraction, ½. The decimal 0.5 is equal to 5/10, and if you simplify the fraction 5/10 by dividing 5 into both the numerator and the denominator, you end up with ½.

Certainly everyone can appreciate the basic concept of fractions but we need to explore further how the addition, subtraction, multiplication and division of fractions is carried out. Now the most basic operation is the addition of two fractions such as ⅛ and ⅝.

Many of you will recall that in adding two or more fractions in which the denominator is the same, one only has to add the numerators together to obtain the correct answer. So if we sally forth and solve the equation ⅛ + ⅝, we find the following: ⅛ + ⅝ = 6/8. We can then simplify 6/8 by dividing both numbers by 2 to get ¾.

No doubt the ease with which we added ⅛ and ⅝ will to tackle an even bigger task—the addition of two fractions having different denominators. Actually the addition of two such fractions is a very simple and straightforward procedure. Suppose we want to add ⅛ and ½. We will immediately notice that the denominators of these two fractions are different. We need not worry because we can solve this problem by finding what is called the "lowest common denominator." This phrase is simply a reference to the conversion of the two different denominators into a single common denominator so that we can make short work of the addition of these two fractions. To add ⅝ + ½, we simply need to find the lowest common denominator among these two fractions. To do so, we must ask the following question: What is the lowest common number by which both of these denominators can be evenly divided? Thinking quickly we realize that 8 is the lowest common denominator because both the 8 in ⅛ and the 2 in ½ can be equally divided into 8. We will then multiply the number of times that the lowest common denominator (8) divides into each of the two denominators in our fraction (8 and 2) by each of the two numerators. As 8 divides into 8 one time, we will then multiply the numerator 1 (in the fraction ⅛) by 1. As 2 divides into 8 four times, we will then multiply the numerator 1 (in the fraction ½) by 4. The net result is that the fraction ⅛ will remain unchanged because the denominator is the same as our lowest common denominator. But the fraction ? will necessarily change to 4/8 because 2 goes into 8 (the lowest common denominator) 4 times; the numerator 1 must then be multiplied by 4 so that the fraction continues to have the same value (½ = 4/8). As we now have the equation ⅛ + 4/8, we need only add the two numerators together to arrive at the answer 5/8.

The subtraction of one fraction from another is similar to the addition of two fractions. The process by which one subtracts fractions will seem very familiar to us due to our having mastered the addition of fractions so easily. First, we need to reacquaint ourselves with the subtraction process. Suppose we have the following equation: ⅔ − ⅓ = ? As with adding fractions with like denominators, we merely need to subtract 1 from 2 to get the answer ⅓. This process is very simple and straightforward because the denominator is the same for both fractions. But what do we do if we need to subtract one fraction from another in which the denominators are different? As with our dissimilar fractions in the addition example above, we need to find the lowest common denominator for both fractions and then subtract the second numerator from the first. If we have the equation ⅓ − ⅙ = ?, we would once again need to find the lowest number into which both denominators (3 and 6) will divide. This number happens to be 6. We will therefore divide 3 into 6, which gives us 2; in turn we will multiply the numerator 1 in ⅓ by 2 to convert this fraction from ⅓ to ²⁄₆. Of course it is still equivalent because ⅓ and ²⁄₆ are the same fraction—the latter is equal to the former; the only difference is cosmetic in that the numerator and denominator in ²⁄₆ have both been multiplied by 2. We can now subtract ⅙ from our newly converted fraction of ²⁄₆ so that we get the following result: ²⁄₆ − ⅙ = ⅙.

For those fractionphiles who believe that fractions are the greatest invention since the number line, there are additional secrets about fractions to be revealed. Now we will shift our focus slightly to the multiplication and division of fractions. Even though the multiplication and division of fractions are typically taught after the addition and subtraction operations, these operations are generally simpler to carry out because they do not require us to go through the trouble of finding a lowest common denominator. In considering the fraction ⅖ × ⅓, for example, we can determine the answer very quickly because all we need to do is to multiply the two numerators (2 and 1) together and the two denominators (5 and 3) together.

Voilà! The product of this effort is $\frac{2}{15}$. The multiplication operation, nonetheless, gives us the opportunity to point out that a very quick way to determine a common (though not necessarily lowest common) denominator when adding or subtracting fractions is to simply multiply the two denominators together. Hence, the number 15 would be the common denominator here if we solved for it.

At this stage, there is only one remaining operation—the division of fractions—that we have not yet mastered. The good news is that the division of two fractions is very similar to the multiplication of fractions, with the inversion of the second fraction as the only added wrinkle. To see how the division of fractions works, we can return to our previous fraction, $\frac{2}{5} \div \frac{1}{3}$, and then invert the $\frac{1}{3}$ so that it is equal to $\frac{3}{1}$ (which, as all fractionphiles in good standing know, is equal to 3). We will now multiply $\frac{2}{5}$ and $\frac{3}{1}$ together to obtain the quotient $\frac{6}{5}$. So $\frac{2}{5} \div \frac{1}{3}$ is equal to $\frac{6}{5}$. If we want to go one additional step, we may express our answer $\frac{6}{5}$ as $1\frac{1}{5}$—which is obtained by dividing the denominator into the numerator (5 goes into 6 one time with a remainder of 1 left over to place over the denominator 5 in the fraction). Accordingly, we have easily mastered what some consider the most intimidating of fractional processes—division .

We should also point out the fractions can also be expressed in terms of positional notation. The term "positional notation" has special meaning for students of fractions because they know that any fraction such as $\frac{1}{4}$, for example, can be expressed as $\frac{25}{100}$. This is really just another way of saying $\frac{2}{10} + \frac{5}{100}$, which can be expressed in the form of a decimal as 0.25. So we see that even the most innocent looking fraction can wear many masks—appearing as a fraction ($\frac{1}{4}$) or in its expanded positional form ($\frac{2}{10} + \frac{5}{100}$) or in its pure decimal form (0.25).

We can appreciate decimals because they are clear and concise and do not take up as much paper as writing out a fraction as a series of fractions highlighting tenths, hundredths, thousandths, ten-thousandths, and so on. But there is a problem that stems from the fact that our world is not as neat and pristine as it might otherwise be. More specifically, our students of mathematics may be distressed to find that not every decimal is so simple and straightforward. Indeed, there are many fractions that, when expressed as decimals, do not end neatly after a digit or two but may go on for thousands, millions, billions, or even trillions of digits without end. An innocuous-looking fraction such as ⅙ does not have a tidy little decimal that sputters out after one or two digits. No, this fraction is far more expansive, with a decimal that begins at 0.16 and then continues on without end boasting an unending sequence of 6s that would stretch to the farthest star of the farthest galaxy of our universe. This unending decimal might be somewhat disturbing because it does not have a clean ending after five, ten, or even fifteen decimal places. But we should not be too dismayed because most fractions have a similar, nonterminating feature. You can carry these sorts of fractions out to any decimal place you may desire and you will still find plenty of numbers looming further ahead in the distance. Even though we may have some philosophical bones to pick with the idea that a nonterminating decimal should exist in the first place, we should console ourselves with the fact that we can carry out such decimal representations to any desired degree of accuracy we may desire. If we need to find out the decimal representation of ⅓, for example, to the thousandths place, we simply divide it and find that it is equal to 0.333. Whether or not we are dealing with a terminating decimal may not make very much difference in the accuracy of many mathematics problems that we may have to consider because we will find ourselves dealing with increasingly smaller quantities (e.g., hundredths, thousandths, ten-thousandths, hundred-thousandths, and so on) that may be disregarded for purposes of many types of calculations.

Because many scientists and mathematicians must deal with very large numbers, they are not necessarily concerned with arriving at an answer that can be determined to the fourth or fifth or sixth decimal place. In calculating the mass of the earth

(6,500,000,000,000,000,000,000 tons or 6 sextillion tons), the number of stars in the Milky Way galaxy (250,000,000,000 or 250 billion), or the number of galaxies in the observable universe (perhaps as much as 1,000,000,000,000 or 1 trillion), for example, scientists must grapple with huge numbers. But because numbers such as ten trillion or fifty quadrillion are so unimaginably vast, we are generally not very concerned about whether we are off by a few thousand or million here and there. In the subatomic world, by contrast, decimals are used to describe the magnitudes of subatomic particles with great precision. The mass of one electron, for example is equal to 0.00000000000000000000000000009 grams—a quantity that is almost inconsequential when compared to a molecule or even a single cell. Protons, by contrast, are comparatively massive, weighing about 0.0000000000000000000000001652 grams. Whether we are off by a decimal place or two or three in the subatomic world can be very significant because that can mean a difference in magnitude of ten, one hundred, or one thousand. This is certainly true when we compare the svelte electron with the bulkier, corpulent proton.

Thus we find that fractions are versatile things that can be added, subtracted, multiplied, and divided to our heart's content. Because the real world is seldom so neat that our numbers divide perfectly evenly, we find that we must often deal with fractional amounts. Hence, we must be ready at a moment's notice to spring into action, pencil and paper in hand, to determine the lowest common denominator. But the fundamental point to bear in mind is that fractions make it possible for us to handle all sorts of quantities that are not perfect even numbers—with great ease and elegance. However, the basic operations of addition, subtraction, multiplication, and division do not exhaust the field of operations that may be used by mathematicians studying the world of fractions. These operations, nonetheless, do form the cornerstone of the subject and must be understood before more sophisticated uses of fractions can be employed.

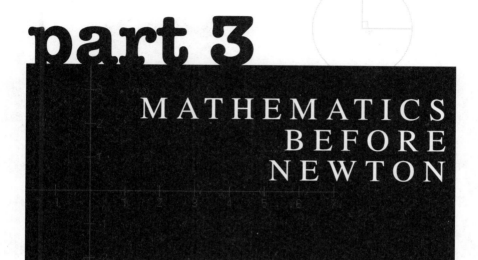

part 3

MATHEMATICS
BEFORE
NEWTON

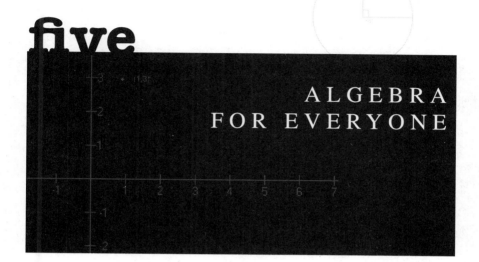

five

ALGEBRA FOR EVERYONE

Algebra is a transitional subject because it is both the capstone of the student's mathematical education at the elementary level and the beginning of his or her study of higher mathematics. Many teachers would not dispute the idea that algebra is the gateway to higher mathematics and that it instills certain skills in conceptualizing the abstract that are essential to further intellectual growth. To learn algebra thus forces one to think like a mathematician and learn to grapple with symbols that may have very profound meanings. Of course many of us do not welcome the prospect of learning a new terminology and discipline but the study of algebra enables us to master a new way of thinking and acquiring knowledge. Once you become comfortable with algebra, you will find yourself better able to think about abstract concepts and to reason in a more logical, clear-headed fashion.

NUMBERS

So where do we begin? Perhaps the best place to start is with that familiar concept of numbers. After all, even the largest oak tree begins as a single acorn, and so must our attempt to tackle algebra begin with its most fundamental component—the concept of number. We are all familiar with the concept of a number line in which a number (e.g., 1, 2, 3, and so on) is marked off at periodic intervals. The most elementary numbers are *natural* numbers such as 1, 71, or 664. You will notice that we used the words "such as" in the previous sentence to let you know that we did not intend to limit the group of natural numbers to only 1, 71, and 664, as this term includes all whole numbers greater than zero. But the number line is far more than a simple repository for the natural numbers since it also includes zero and all negative integers (e.g., –1, –2, –3, and so on).

But there is more to the number line than simply positive and negative whole numbers (integers). It can also claim as its own all nonintegers (fractions) such as ⅔, –⁴/₁₇, or ⅗. This suggests that the number line is much "denser" than we may have previously thought. There are an infinite number of such non-integers between every two integers on the number line. Although this concept may at first be difficult to fathom, we can retrieve our trusty number line and look at the interval between 1 and 2 for a better understanding. There are many fractions that, when reduced to their lowest possible terms, could be placed between these two points. Examples would include ³/₂, ⁵/₄, and ⁸/₇. By altering the numerators and denominators accordingly, we could devise an unlimited number of fractions that we could insert between 1 and 2 on the number line; this fact demonstrates that the number line is more complex than we may have believed.

The number line also includes all negative fractions such as –⁴/₅, –⅐, and –⅗ as well as all decimal expressions like 1.56, –1.99, and –1.14. But our catalog of decimal expressions is much more extensive than even these illustrations would suggest as it can include decimals that are only a tenth, hundredth, thousandth, or even less of a whole number, such as 1.1, 1.67, and 1.453. Indeed, one can create decimals as small as one likes merely by extending the sequence of numbers further and further to the right of the decimal point. So we can certainly have decimals of the magnitude of hundred thousandths, millionths, billionths, or even trillionths, all of which can be positive or negative. Moreover, these decimals can be tucked in between any two whole numbers depending on the quantities we place to the left of these decimal points: 1.07, −1.866, 1.9876, and so on. This variety of expressions can be found throughout the number line, along both its positive and negative sides.

No doubt you may already feel that you have learned everything there is to know about the number line, but you will be pleasantly surprised to learn that we have only glimpsed the tip of the iceberg. To this point, we have discussed those numbers that can be characterized as *rational* numbers in that they can all be expressed exactly as a ratio of two integers (or, alternatively stated, one integer divided evenly by another). But the number line also boasts *irrational* numbers such as $\sqrt{8}$ and $3\sqrt{11}$ that, not surprisingly, cannot be expressed exactly as a ratio of two integers. Both rational and irrational numbers make up the group of *real* numbers, constituting the entire class of numbers that can occupy distinct points on the number line. But the number line is not the alpha and the omega of mathematics because it does not contain a very peculiar class of numbers known as *imaginary* numbers. What are imaginary numbers? The term is confusing and perhaps unfortunate because we think of all numbers as being imaginary or at least conceptual in nature. But imaginary numbers are the square roots of negative numbers such as $-\sqrt{5}$, $-\sqrt{9}$, and $-6\sqrt{3}$. Try as we might, we cannot place these numbers on the number line. But it is useful to know that the real numbers and the imaginary numbers together constitute the set of *complex* numbers.

You now know more about the number line than likely 99 percent of the general population. The number line, in short, is the graphical representation of the set of real numbers and a subset of the set of complex numbers. It contains several subsets of numbers, including all positive and negative numbers that in turn include all rational and irrational numbers that in turn contain integers, nonintegers, *radicals*, and so-called *transcendental* numbers (those numbers, such as pi, that cannot be expressed as the roots of integers).

The purpose of this digression is to illustrate the basic structure of the number system. But the definition of various types of numbers is but one part of the story. To simply recite a litany of different types of numbers—whether they are integers or radicals—does little good if one has no sets of rules for governing the use of these numbers. It is analogous to being given a set of checkers and a board without having any idea as to how to play the game itself. Our being able to put the checker board and pieces to practical use would necessitate that we become familiar with the rules of checkers. Once we learn about the objective of the game (taking all of the opposing pieces) and the actual mechanics of moving the pieces including jumps, then we can actually play a game of checkers.

We need to know the rules of whatever game we are playing in mathematics in order to be able to utilize the numbers in a meaningful way. So what are the rules of algebra? Although the rules that govern algebraic operations are somewhat numerous and detailed, we need to be cognizant of some of the most important axioms that apply to the multiplication and addition of numbers. "Axioms are propositions which are laid down without proof, in order to prove further propositions (theorems) from them."[1] These axioms— which are known collectively as the field axioms—include (1) closure, (2) commutativity, (3) associativity, (4) distributivity, (5) identity elements, and (6) inverses. Any set of numbers that obeys all of these axioms is known—quite cleverly—as a field. So it now remains for us to briefly examine each of these axioms in turn, if only to lay the groundwork for our future mastery of the subject.

MEET THE FIELD AXIOMS

What is closure? Although it is perhaps one of the most overused psychological terms of the decade, pertaining to resolving unfinished issues of a personal nature (such as a relationship with a loved one or a spouse), it also has a very important place in mathematics. When we say that a set is closed, we mean that we cannot get an answer not contained within the set when we perform a mathematical operation such as multiplication or addition on the set. Suppose we have the set $(-1, 0, 1)$. We would say that this set is closed under multiplication because all of the answers that we obtain from multiplying these numbers are found within this set of numbers. For example, $0 \times 0 = 0$; $0 \times 1 = 0$; $0 \times -1 = 0$; $1 \times 1 = 1$; $-1 \times -1 = 1$; and $1 \times -1 = -1$. If we could carry out all of the different possible combinations of multiplicative operations in which 0, -1, and 1 can be used, then we would see something amazing take place: We would find that every answer we can obtain by multiplying any of the elements in this set is found in the set. But we also see that this very same set is not closed under addition because if we take the equation $1 + 1 = 2$, we find after checking the three elements in our set that 2 is not one of those three elements.

We will expand our knowledge of algebra by considering the axiom of commutativity. In mathematics, commutativity means that we can exchange the positions of numbers in equations and not alter the outcome of those equations. Addition is commutative because $2 + 5 = 5 + 2$. The fact that we switch the positions of the two numbers does not alter the fact that both numbers, when added together, equal 7. The same can be said for multiplication because $2 \times 5 = 5 \times 2$. Even if we are particularly determined to disprove this axiom and carry out the same operations twenty or thirty thousand times, we would find that the result would always be the same. But not everything in the world is commutative—which is unfortunate for those of us who admire symmetry.

The axiom of associativity merely refers to varied groupings of numbers that, at least in the case of addition and multiplication, remain equivalent to each other, even as their positions in the equation are swapped. So $5 + 3 + 4 = 4 + 5 + 3$. But this equivalence of the axiom of associativity does not hold true when we consider subtraction or division. We could take the equation $4 - 1 - 2 = 1$, which is not equal to $2 - 1 - 4 = -3$ even though we are dealing with the same three numbers. The only thing that has changed is the position of the numbers. The same inequality would appear if we were to substitute a division sign for the subtraction sign because the two expressions would not be equal.

Our introduction to the fascinating world of multisyllabic algebra continues with the axiom of distributivity, which merely permits us to spread out an equation such as $3 \times (2 + 4)$ and construct its equivalent expression: $(3 \times 2) + (3 \times 4)$. But the axiom of distributivity does not hold up if the original equation is changed to $3 \times (2 \times 4) = 24$ because if we distribute the equation we get $(3 \times 2) \times (3 \times 4) = 60$. So its application is even more limited than that of associativity or commutativity.

We now move on to two more intriguing axioms—identity elements and inverses. The numbers 0 and 1 are identity elements because any number multiplied by 0 is equal to 0 and any number multiplied by 1 is equal to that number. The identity axiom may be easily remembered if we consider that the word "identity" implies something equals itself. Inverses are somewhat similar in principle because any number that is multiplied by its inverse (for example, ⅓ is the inverse of 3 and ⅕ is the inverse of 5) is equal to 1. These axioms are wonderful tools that mathematicians can use because they apply to all numbers. We can instantly determine the inverse of any number because it is not a difficult exercise to place a 1 over that number. Having done so, we can then triumphantly proclaim, for example, that 1,000,000,000,000,000 multiplied by its inverse is equal to 1.

VARIABLES

Having learned about the field axioms, we now should discuss the variable, which is one of the simplest concepts in algebra and, paradoxically, perhaps the one most responsible for the negative image of the subject itself. A variable is any non-numerical symbol—usually an alphabetic symbol such as the letter x or y—appearing in an expression. A variable is nothing more than a number clothed in alphabetic garments, but its very presence intimidates some people because of its aura of mystery. It may be more helpful to think of a variable as someone who likes to play dress-up games. Sometimes the variable takes on the identity of a 2 or a 4 and other times it adopts the identity of a 6 or a 9. But the important thing to remember is that we must be consistent in the way we assign values to specific variables. In other words, if we have an equation such as $4x^2 + 2x + 2 = 14$, we will typically be asked to find the value for x that makes this equation true. Now we might hazard a guess that we can solve this equation if we assign the value of 2 to x. We would find that this equation is indeed true if we substitute 2 for x. But we must substitute the same value for x at every single place x appears in the equation because an equation must have its own internal consistency to be mathematically valid. It would also be impossible to solve the equation in which x did not have the same value throughout because any solution of an unknown variable such as x depends on our being able to shove all the unknown variables to one side of the equation and all the specific numbers to the other side. In this way, we can express the variable x in a single defined term. To return to our equation $4x^2 + 2x + 2 = 14$, our algebra would be virtually worthless if the values for x could change from term to term in a single equation. Imagine that the x in $4x^2$ was equal to 3 and the x in $2x$ were equal to 5. Certainly we would have very little idea as to how we should proceed to solve the equation because we would not know which values had been assigned to which variables prior to our considering the problem. So we must be consistent: if

x is equal to 4, for example, in one term, it must be equal to 4 in all the terms of the equation. In short, it must represent the same number being considered at any given time in the same equation. But this certainly does not preclude the x variable from taking on a completely different numerical value in a different equation at a different time. Perhaps it is best to think of our variables as chameleons in that they can take on any one of an endless variety of numerical values but any single variable can only take on a single value in a given equation. Today x might be equal to 2; tomorrow it might be equal to 4. The changes in the values assigned to x will be dictated by the equation itself but these changes must always be consistent. The fact is that mathematics is the most rational—or at least the most logical—of all the sciences because its integrity is completely dependent on its own self-consistency. It is valuable because its principles are immutable and universal in their application.

ORDERS OF OPERATIONS

A mastery of mathematics depends on knowing the rules relating to the order of its operations. We are referring to the order in which we carry out mathematical operations in an equation. What do we mean by "operations"? Quite simply, we are talking about arithmetical operations such as addition, subtraction, multiplication, and division. But why should it make any difference as to the order we perform these operations? A look at a simple equation may provide us with some insight into the importance of these rules. Suppose we have the expression $5 \times 4 + 6 - 3 = ?$. There are several ways we could solve this equation. We could multiply 5×4, add 6, and then divide by 3, which would give us 23. Or we could add 4 and 6, subtract 3, and then multiply it by 5, which would give us 35. Or we could add 4 and 6, multiply it by 5, and subtract 3 which would give us 47. We have carried out these mathematical operations in three different orders, and have obtained three different answers for the same equation. For those of us who recoil from the unflinching rigors of the tradition-laden mathematics that we have come to know during our educational careers, this new, more "flexible" mathematics might have a certain appeal. But our sense of logic would prevail. The problem with utilizing a more flexible approach is that it essentially destroys the usefulness of the subject itself. After all, a virtue of mathematics is that it boasts a set of rules that give it predictability and immutability. The equation $2 + 3$ is equal to 5, regardless of whether we add the two numbers together now or five hundred years from now. Similarly, the rules of mathematics do not change with shifts in geographical location. Thus we might say that mathematics is "geographically invariant," which is a fancy way of saying that our equation is going to have the same answer regardless of whether we are solving it on a paradise island or in an alpine village or even while standing under a waterfall. Mathematics is fundamentally rooted in the idea that the same equation will yield the same result for all time. This invariability is made possible by a hierarchy of rules relating to the different mathematical operations that provides us with an unchanging order in which these operations are to be carried out: First, we perform any operations that are contained within parentheses. If there are more than one set of parentheses (e.g., parentheses within brackets), then we need to begin with the innermost set of parentheses and work our way outward. Second, we perform any exponential operations. Third, we carry out multiplication and division operations in the order in which they appear (left to right). Fourth, we perform addition and subtraction operations in the order in which they appear (left to right). So if we are given an equation such as $2 + (2 \times 5)^2 - 6$, then we will first multiply the amount within the parentheses (which is 10). Next, we will raise 10 to the second power (10×10), which will give us 100. Following the logic of our trusty rules, we will then add 2 (which is equal to 102) and subtract 6, thus leaving us with a final answer of 96. We can see that we could have come up with any number of other answers had we chosen to solve this equation using a different order of operations. But if we did not have a single set of rules to govern the order of operations, then we would have a sort of Tower of Babel in mathematics with every-body speaking in a different tongue or at least coming up with different answers for the same equations. Needless to say, our mathematics would not be very useful if it could not be universally replicated. So these rules regarding the order of operations are among the most important features of mathematics.

Simplicity in Algebra

Although the exotic appearances of certain algebraic equations may cause us to believe that mathematics is a competition to see who can devise the most forbidding, complicated equations, you might be surprised to find that mathematicians, by and large, are driven by a desire to simplify these equations. The primary reason for ex-pressing algebraic equations in less cumbersome forms is that they are easier to solve. There is also not very much reason for mathe-maticians to complicate their mathematical work intentionally because they simply do not need to deal with an additional layer or two of complexity in their calculations. Most of us would subscribe to the notion that simplicity is a good thing—even in algebra.

But how does this notion of simplicity manifest in the mathe-matical world? We see it implicitly when we have an equation such as $4x + 5 = 5x - 13$. The first thing any enterprising problem solver does, as we noted above, is to move all of the variables (unknowns) to one side of the equation and all of the numbers (knowns) to the other side of the equation. We thus obtain the equation $4x - 5x = 13 - 5$, which, when simplified, reads as follows: $-x = 8$. Although this grouping may be prompted in part by an aversion to disorder and a general enthusiasm for neatness, it also results in a simplification of the equation that makes it easier to solve. So we need to keep in mind the idea that like terms need to be placed together to the extent we can do so to facilitate our search for an answer.

Another tool used for simplifying algebra is the concept of absolute values. The absolute value of a number refers to its position on a number line relative to the origin (point 0). "The absolute value of a real number a, written $|a|$ (read: mod a), equals a if a is positive, and $-a$ if a is negative, e.g., $|-2| = 2$, $|+2| = 2$."[2] So the absolute value of 5 and -5 are both equal to 5 because we are concerned here simply with the number of points (each point representing a whole number) one would have to move along the number line from 0 to get to that particular number. In the case of 5 and -5, we would obviously have to move 5 points to the left or the right of 0. The amount of displacement is thus equal to 5 units on this number line.

Breaking Down Equations

To solve complicated mathematical expressions requires a certain burrowing skill depending on knowledge of the order of operations. In short, we need to be able to dig ourselves out of the most complex parentheses-laden equation by tunneling from the innermost terms outward. If, for example, we have the equation $[2 + (4 + [8 \times (5 + 4)] - 3)]$, where would one even begin? We must go back to the rules of operation and recall that we must solve the innermost equation first and work our way—or, if you will, tunnel our way—outward. Here, we first begin by moving inward to the innermost equation. This may not refer to the most centrally located term, in a geographical sense, in the equation, because the varied arrangements of brackets and parentheses may be such that the innermost equation and the one that is located in the middle of the expression are different. The foolproof way to approach this search for the innermost equation is to locate each bracket and parentheses and its corresponding partner and then move inward until reaching the final pair of parentheses or brackets. Your sweep through will yield the term $(5 + 4)$ because it is flanked by two sets of brackets and another set of parentheses. To solve this equation, we would begin with the term $(5 + 4)$ which is equal to 9. We would then move outward to the first set of brackets, which would entail multiplying 9 by 8. Having obtained the product 72, we would then add 4 and subtract 3 (moving left to right) to obtain the sum of 73. Finally we would add 2 in moving to the outermost set of brackets to obtain the answer 75. This approach illustrates that we can take the most forbidding-looking mathematical expression and, by keeping our composure, reduce it to a series of steps that can be as easily handled. Indeed, that is the key point to remember about all mathematical equations: They should not be viewed as single equations but instead as a series of smaller equations that can be broken down and solved.

POLYNOMIALS

Any tour of the world of algebra must involve a look at polynomials. Polynomials are merely algebraic expressions that are used in addition, subtraction, and multiplication operations. Although this may sound like a limited repertoire, it actually includes quite a diverse group with an endless variety of manifestations. So algebraic expressions as different as xy^5z^3 and $x - 6$ and $6y$ and $8x^5 + 3x + 1$ are all polynomials. Polynomials include all real numbers and, by definition, exclude all imaginary numbers (e.g., $5\sqrt{3}$, $-\sqrt{10}$). So any expression that requires us to divide a number by a variable such as $\%$ is not a polynomial. Similarly, the taking of a square root of a number less than 0 such as -1 would give us an imaginary number that would lie outside the realm of the polynomials.

The word "polynomial" suggests a group of some sort. Indeed, we would not be surprised to find that the "polynomial" family consists of a variety of different entities including "monomials," "binomials," and "trinomials." Actually the only difference between a monomial and a binomial and a trinomial is the number of terms in the expression. For those of you who studied your ancient languages, you know that a monomial must consist of a single term. So the expression $3xy^4z^3$ would be considered a monomial because it is a distinct term even though it consists of several different variables that could be multiplied together. But the expression $3xy^4z^3 + 5yz^3$ would be a binomial because it would involve two terms linked by a "+" sign. A trinomial, as you would guess, consists of three terms as in the equation $3xy^4z^3 + 5yz^3 - 2x^2$. By determining the number of terms to be 3, we can see why the name "trinomial" would be appropriate. But what about expressions having even greater numbers of terms? Although there is no commonly accepted nomenclature, we could use the term "quadnomial" to refer to an expression having four terms such as $3xy^4z^3 + 5yz^3 - 2x^2 \times 8y$. Of course the crucial point here is the number of terms themselves and not the sheer number of variables within each term.

Polynomials are very flexible mathematical equations and, in the finest algebraic tradition, can be manipulated from one form to another. We can have two trinomials, for example, which, when added together, yield a monomial. How can this apparent sleight of hand be carried out? Suppose that we have $(8x^3 + 3x + 6) + (x^3 - 3x - 6)$. We shall see that $8x^3 + x^3$ is equal to $9x^3$. But the other two terms in each equation cancel each other out because $3x - 3x$ is equal to 0 and $6 - 6$ is equal to 0. Sure there is no point in stringing a couple of 0s after $9x^3$, we essentially are left with a monomial term even though we started out with a total of six terms in two separate trinomial expressions. Certainly this is an impressive feat that also enables us to simplify the outcome of our arithmetic operation.

Closely related to the concept of polynomials are *factors*, which are essentially the building blocks of polynomials. Suffice to say, every polynomial has one or more factors. But what is a factor? It is those parts of the polynomial that are multiplied together to create the polynomial. In the polynomial $6xy$, for example, there are three factors (6, x, and y) that are multiplied together to create $6xy$.

There is nothing particularly remarkable about the polynomial $6xy$. But it should be noted that the number of variables that are multiplied together determines whether it is a second degree polynomial (as is the case with $6xy$) or a fourth degree polynomial (like $3x^2y^2$) or some other degreed polynomial. This concept of degree offers a shorthand reference to the number of variables that are being multiplied together to create a given polynomial (they do not include the "3" in this equation). When we refer to the number of variables, we are not talking about the number of distinct variables (e.g., x, y, z) but instead the number of variables themselves (e.g., x^2 [2 variables], y^3 [3 variables], etc.). So if we were to analyze the polynomial $5x^2y^3z$, we would find that it consists of two x variables, three y variables and one z variable multiplied together. It would thus be known as a sixth-degree polynomial because it would consist of a total of six variables, even though there would only be three different letter variables in this polynomial.

To be absolutely clear as to the nature of the nomenclature that is being used here, let us look at more examples. The variable $4x^3y^2$ is a fifth-degree polynomial $(x \cdot x \cdot x \cdot y \cdot y)$, whereas x^2y is a third-degree polynomial $(x \cdot x \cdot y)$. What if we have no variables at all? No x variables nor y variables nor z variables? You may think that we have stumbled upon one of those internal contradictions that could bring the entire edifice of mathematics crashing down all around but any number that lacks a variable is not a polynomial at all but a constant. A number such as 32 clearly lacks a variable. So its value cannot change as the value of x or y changes. The constant is unchanging in its demeanor. It will always have the same numerical value. So we call such constants—whether they are 32 or 12 or 65 or 124—zero degree expressions. They are like the proverbial Rock of Gibraltar and will forever remain the same, sturdy and unflinching.

Let us further familiarize ourselves with the different pieces that make up a typical variable. If we take $4x^3$, we quickly see that it has several distinct parts. The 4 is the coefficient, the x is the base, and the 3 is the exponent. The exponent 3 tells us that we must raise x to the third power or, alternatively stated, multiply $x \cdot x \cdot x$.

When discussing polynomials, one of the things that comes to mind is the need for a quick and straightforward process for multiplying polynomials. In general, we are most likely to be faced with the task of multiplying two binomials such as $(x + 3)$ and $(2x - 2)$. Although this may appear to be a daunting task, we only need to remember that each term in the first binomial must be multiplied by each term in the second equation. So we would take our x variable and multiply it by $2x$ and then by 2, which would give us $2x^2 - 2x$. Well, that was simple. But we are only half-finished as we must now multiply 3 by those very same numbers which will give us $6x - 6$. The resulting product is $2x^2 - 2x + 6x - 6$ which can be simplified as $2x^2 + 4x - 6$. This trinomial is a straightforward example of the ease with which binomials may be multiplied.

Even though multiplying binomials may not be your cup of tea, it is conceptually easier than factoring a polynomial. Multiplying two binomials merely requires that one engage in a series of arithmetic calculations. Factoring polynomials, by contrast, is not necessarily such a straightforward process because if we were to take the trinomial we obtained previously, we would see that $2x^2 + 4x - 6$ does not readily show which two binomials can serve as its factors. We know that $2x^2$ can be broken down into $2x$ and x and so we can easily obtain the first term in each of the two binomial factors. But then we have to find two additional numbers that will, when added together, equal $4x$ and will, when multiplied together, equal -6. This often entails a little bit of guesswork. The most efficient approach is to try to find pairs of numbers that equal -6 and then see which of those pairs, when added together, equal $4x$. After some trial and error we will eventually discover that the pair $(3, -2)$ will fill satisfy our requirements very nicely. The resolution of this equation will also force us to recall the basic rules governing the multiplication of negative and positive numbers: If we multiply either two positive or two negative numbers, the product will be positive. But if we multiply one or more positive numbers with an odd number of negative numbers, then the product will also be negative.

Sometimes we will be fortunate to be asked to solve equations that are already factored and set equal to 0. One example would be $(2x + 6)(4x + 2) = 0$. Now there is actually a lot of information contained within this equation that we merely have to extract. We know that the product of these two equations equals 0 because that is information given to us by the equation itself. Moreover, we also know that $2x + 6 = 0$ and $4x + 2 = 0$ because one of the two factors must be equal to 0 for the product of these two equations to be equal to 0. This is a very basic axiom of mathematics. We can find the solutions to x for each of these two problems fairly easily. If $2x + 6 = 0$, we know that $x = -3$. How? We return to our technique of grouping variables on one side and numbers on the other to solve the unknown value—in this case, the variable x. We first subtract 6 from both sides of the equation to get $2x = -6$. Then we divide both sides by 2 which gives us $x = -3$. In the same way, we know that if $4x + 2 = 0$, x must be equal to $-\frac{1}{2}$. So our solutions for our original equation would be $(-3, -\frac{1}{2})$. But we may have to take this analysis a step further. If the domain of the variable (x) were equal to the set of integers, then -3 for x would be a solution because -3 is an integer. But $-\frac{1}{2}$ could not be a solution because it is not an integer. Of course if $-\frac{1}{2}$ were actually some integer that was merely being expressed in a fractional form such as $-\frac{4}{2}$ (for 2) then it would be an integer and could be considered a solution to the problem. But if the domain of the variable were the set of all natural numbers (positive whole numbers), then the solution set for the expression $(2x + 6)(4x + 2) = 0$ is an empty set because there are no solutions that are natural numbers.

Mathematicians must also grapple with extraneous solutions to equations that rear their heads when the original equation is manipulated in a way (such as multiplying it by a variable) to create a new transformed equation. In this new (though not necessarily improved) version of the original equation, any solution to the original equation should still work for the new equation. But, interestingly, by transforming this equation through various algebraic manipulations, we may also obtain solutions that work for the transformed equations but do not satisfy the original equation. These solutions are considered to be extraneous to the original equation. Here as with everything else in life, an example will help to clear the air. Suppose that you are sitting for your doctoral dissertation defense in mathematics and the committee of four professors conducting the examination decides to give you a single equation to solve. They slide a piece of paper over to you which you pick up, knowing full well that your entire academic future depends on your being able to answer this problem. Imagine your relief when you learn that you must solve the following equation: $x = 2$. Now even a young child could figure out that the solution for this equation is equal to 2 because $2 = 2$. But if you should decide to impress the dissertation committee by manipulating the equation by multiplying both sides by $x - 4$, then you would obtain the following transformed equation. $x(x - 4) = 2(x - 4)$. In solving this new equation, we would see that it has an additional solution of 4 that could not be obtained in the original equation $x = 2$. But this venture into the realm of algebraic manipulation does not provide us with any additional solutions for the original equation. In other words, the transformed equation yields what is known as an extraneous solution that does not satisfy the original equation so it will really not impress the committee. Moreover, you cannot always undo the transformation of the original equation by trying to divide it by the same expression that you used to multiply it—at least in cases in which the expression is equal to 0; since dividing a number by 0 is undefined. Hence, when we multiply both sides of an equation by an expression that is equal to 0, we are taking an irreversible step that cannot be undone mathematically to bring us back to the original equation—no matter how much we beg and plea and whine to the review committee.

INEQUALITIES

Many of us remember our early grade school years when we were first introduced to the inequality signs used to state that one side of an equation is greater than ("$>$") or less than ("$<$") the other side of the equation. When asked to solve inequalities such as $7x - 4 > 24$, we first need to avoid panicking merely because we have an inequality sign instead of the tried and true equivalency sign. Here, we can proceed in the same way that we normally would if we had the familiar equivalency sign. We would add 4 to the other side of the expression in order to simplify it. Thus we would have $7x > 28$, which requires us to find all values of x that make this statement true. We could substitute the value of 4 for x, which would give us the following, incorrect statement: $7(4) > 28$ or $28 > 28$. This statement is obviously not true because 28 cannot be greater than 28 as they are both the same number. But we would soon deduce that if we substitute any number greater than 4 for x, we will obtain a true statement. So 5, 6, 7, 8, 9, . . . and so on can be substituted for x and can be used to make our statement of inequality valid. Here, we see that instead of one or two solutions, we have an infinite number of solutions because any number greater than 4 when substituted for x will satisfy the expression $7x > 28$. If you are a person who likes having only a single solution for a given problem, then you may find that inequalities will leave you with feelings of uncertainty. They have a certain open-endedness that can be a little disconcerting at first, but eventually you might revel in the variety that is offered.

If we are visually oriented, we could graph inequalities using a number line. In the inequality $7x > 28$, we could graphically describe the solution by drawing a number line that is marked with whole numbers. We can then show the solution for this inequality by drawing a bold line that begins at 4 and continues moving toward the right indefinitely to include 5, 6, 7, 8, and so (see figure 1).

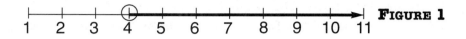

FIGURE 1

Since 4 itself is not a number that would make this statement true, we place an open zero at the 4 to indicate that it is not part of this solution set for this particular inequality. If our inequality was $7x \geq 28$, however, we would look for solutions that, in this case, are greater than or equivalent to 4. Then we would graph the solution by placing a closed circle over the 4 on the number line and continue the bold line toward the right through 4, 6, 7, 8, and so on. The bold line would tell us to include the endpoint of the line (in this case, 4) in our illustration of the solutions of the equation.

Now that we have warmed up to the task at hand, we can take this issue of inequalities one step further. What should we do if we get an inequality such as $-4x \geq 16$? As with most things in life, we want to simplify this equation before we try to solve it. So we need to get rid of the negative sign and find the value for x. Any mathematician worth his or her salt will multiply $-4x \geq 16$ by $-\frac{1}{4}$ on both sides to get the equation $x \geq -4$. However, this expression is not the same as our original expression because our multiplying both sides of the equation by a negative number has completely reversed the order of the numbers. We now have a positive number on the left side and a negative number on the right, whereas we began this particular excursion with a negative number on the left and a positive number on the right. So what we have to keep in mind is that any time we multiply each term in an inequality, we need to reverse the inequality sign or, to use the jargon of mathematicians, "flip it" so that our equation is expressed as $x < -4$. Thus our solution set for our original equation $-4x \geq 16$ includes the numbers -4, -5, -6, and so on. Our graph of this solution would be a black line beginning with a closed circle at -4 and then moving toward the left indefinitely to include all negative numbers.

Suppose we have $|x| > 16$ and are asked to graph the solution set. We recall that the $|x|$ refers to the number of units or, alterna-

tively, the distance from the origin 0 to x on a number line. The "| |"
tells us to disregard whether a number is negative or positive and
instead seek its absolute value. In the case of $|x| > 16$, we would be
including any numbers greater than 16 and less than -16 as shown
in figure 2.

FIGURE 2

Our graph of the solution has two components. The first is a line
that begins with an open circle at 16 (to show that 16 itself is not a
member of the solution set), and continues upward without limit
through 17, 18, 19, and so on. The second line also begins with an
open circle, but at -16; it then continues downward without limit
through -17, -18, -19, and so on.

What if we were to get carried away and reverse the inequality
sign to give us the equation $|x| < 16$? We would then determine that
our solution to the equation $|x| < 16$ includes every number between
-16 and 16 as shown in figure 3. Here we draw a segment on our
number line that is bounded at both ends (-16 and 16) by open circles to show us that our solution set is those numbers between these
endpoints including -15, -14, -13, . . . , 13, 14, 15.

FIGURE 3

Consistency is a virtue in mathematics. We spoke earlier of the
need for the value of any variable in an equation to have the same
value throughout that equation so that it would be possible to solve
the equation. That the value for the variable x in a given equation
will remain consistent throughout that equation is required by the
"reflexive" property of mathematics, which says that if x is a real
number, then $x = x$. Such a statement will not strike one as being particularly profound because it appears to be self-evident: $x = x$. But

the importance of this axiom is that it reminds us not to allow our variables to assume different values within the same equation. Although this axiom may appear to be simplistic, it must be followed so that the internal consistency of our mathematics can be maintained. Otherwise, we would not be able to solve any mathematical equation with any confidence because we would never know what values might be assumed by our chameleonlike variables.

Closely related to the "reflexive" property of mathematics is the idea that mathematics is symmetrical; this idea is underscored by the "symmetry" axiom. This axiom tells us that if $x = y$, then $y = x$. Here again, we are dealing with a statement that seems to be patently obvious to us, but it is one that also insures the logical consistency of our mathematics. We would have quite a chaotic situation if x could be less than, equal to, or greater than y, all at the same time in that same equation.

Let us now explore the transitivity axiom, which bolsters the logical consistency of mathematics by offering a sort of multistep reasoning process whereby if $x > y$ and $y > z$, then $x > z$. This axiom is more profound than it might first appear. It makes it possible for us to extrapolate our reasoning process so that we can make a statement about the comparative values of two variables even though we may have no direct evidence as to how these two values compare with each other. The transitivity axiom may be understood by comparing the heights of three different individuals. Suppose Ed is very tall and always bumps his head when going in and out of his front door. But Ed's neighbor, Stan, is even taller than Ed and smacks his entire face against the door frame when visiting Ed's house. So we can express the fact that Stan is taller than Ed by use of our own shorthand algebra notation: Stan > Ed. But suppose that Ed's other neighbor, Charles, is taller than Stan and bumps his shoulders when attempting to pass through Ed's front door. We can say that Charles is taller than Stan by using our notation a second time: Charles > Stan. Now we have two quantitative relationships that we have expressed using inequality signs. But the transitivity axiom makes it possible for us to relate these two inequalities by offering a third inequality relating Ed's height to that of Charles's height. So our transitivity axiom permits us to utilize the following reasoning: If Stan > Ed and Charles > Stan, then Charles > Ed. Even though Charles and Ed have never been seen together in public, the transitivity axiom still makes it possible for us to make a statement about their relative heights because we know how each of their heights compares to our common denominator—Stan. But this same line of reasoning would apply if Ed, Stan, and Charles were all the same height. Using our algebraic notation, we would express the heights of the three individuals in the following way. If Stan = Ed, and Charles = Stan, then Charles = Ed.

The transitivity axiom leads us to its second cousin, the trichotomy axiom, which states that if we have any two real numbers x and y, then one and only one of the following three relationships can be true: $x > y$, $x < y$, or $x = y$. The beauty of the trichotomy axiom is that it is all-inclusive: a number is either less than, equal to, or greater than a second number. There is no room for the wishy-washiness that is the refuge of the muddled thinker. Some people might find this absolutism to be symptomatic of a tyrannical dictatorship while others might be comforted by the fact that the range of choices in the trichotomy axiom are so few and distinct.

What is the point of these axioms? Quite simply, they provide all of us with the tools to venture into the wilds of the mathematical terrain and construct logical proofs that in turn make it possible to develop entire branches of mathematics. It may be helpful to think of these axioms as the "engine parts" of the mathematical machine. These axioms must hold true for the whole mathematical edifice to stand.

MATHEMATICAL REASONING

The focus of mathematics—unlike branches of the applied physical sciences—is primarily on abstract concepts. This may be frustrating to those hardy types who disdain Descartes' pithy declaration, "I think, therefore I am!" because they want something to touch and smell and see and taste. Mathematicians, alas, must rely on the theoretical validity provided by their subject's logical structure. Pure thought—not rockets or chemical substances—is the mechanism that underpins mathematics. And few things have had a greater impact on the field of mathematics than the development of the hypothetical statement, which we usually see manifested in an "If, then" form.

What do we mean by an "If, then" statement? Quite simply, it is any statement that relies on the validity of certain basic premises (the "if" part) for determining the validity of the conclusion (the "then" part). Perhaps the easiest way to proceed is to take a typical "If, then" statement that appears in all of the leading mathematics textbooks: "If Michael is a plumber and all plumbers are members of the Blades card club, then Michael is a member of the Blades card club." Now we have no idea as to whether Michael is a competent plumber. But we are not really interested in his professional expertise. After all, we want to understand better how this logical reasoning process works. The "if" part of the sentence relating to Michael and his membership in the Blades card club is the premise of this statement. The premise is an arbitrary statement that provides a sort of beginning point in our reasoning process. From the premise we then move forward to the "then" part of our statement that constitutes the conclusion of the statement. The premise contains two or more statements of fact. The conclusion, however, is drawn from the premise and so its validity is totally dependent on the premise. To return to our example, we would take as a given that Michael is a plumber and that all plumbers belong to Blades. We would be able to conclude from the premises of our statement that Michael belongs to Blades because he is a member of a class of individuals who all belong to Blades. So we would not need to verify further his membership in Blades.

The beauty of this type of reasoning—which logicians call

deductive reasoning—is that it permits us to draw general conclusions from specific statements of fact. We are able to make sweeping generalizations based upon comparatively few observations. In the case of Michael and his fellow plumbers, we do not have to check the membership rolls of Blades club or poll all of the plumbers to verify that they enjoy spending their evenings around the card tables. We merely have to check the logical consistency of the premises of our statement to see whether the conclusion does indeed flow from these assumptions.

The use of deductive reasoning in mathematics is necessary to avoid the evidentiary trap inherent in inductive reasoning. What is inductive reasoning? It is an intellectual process whereby the deep thinker makes numerous observations in the world about him and then forms conclusions based on those observations. This sounds like a straightforward process but it can snare the unwary into a trap. Why? The inductive reasoning process requires one to draw general conclusions based upon specific observations.

This point can be illustrated by following the exploits of a zoologist who has spent much of his life wandering through the jungles of the world studying the fur shades of the lion. After thirty years of observations involving more than two thousand lions, our zoologist has concluded that the fur patterns of the lion come in four distinct shades of brown. Our zoologist has never seen a lion with black fur so he has gradually concluded that all lions have one of these four distinct shades of brown fur. But our zoologist can never be completely certain that there is not a lion somewhere in the world with a black coat. He can only tentatively conclude that lions have four shades of brown fur. His conclusion, however, does not inevitably follow from his collections of observations because of this lack of complete certainty. A major shortcoming of inductive reasoning is that even in cases in which a person makes hundreds or even thousands of observations, that person cannot say that the conclusions he draws from the observations will always hold true in all cases. Instead, someone such as our zoologist can only offer a best guess as to the true state of affairs regarding the shades of color of lion fur. His conclusion may be a very good guess as no one has ever seen any other colored fur on a lion, but he cannot say with complete certainty that this is the true state of affairs among lions. He will never be able to say for certain that there are only four shades of brown fur on lions because of the slim chance that there is a different colored lion wandering around that does not fit into one of these categories. Our zoologist would doubtless be much happier if he could count upon the certainty of deductive reasoning.

Deductive reasoning is particularly well suited for the world of mathematics because it does not necessarily look to the real world for validation. The point of the deductive reasoning process is not to verify the premises("if") of the statement but to see if the conclusion ("then") drawn from those premises makes logical sense. So our zoologist could enjoy a very deep and restful sleep if he were able to construct a theory about the colors of lion fur using deductive reasoning. One type of deductive statement would be the following: "If all lions have brown fur and that animal has black fur, then that animal is not a lion." This statement is logically consistent because it suggests lions as a class only have brown fur and invariably leads us to the conclusion that any animal with black fur is not a lion. This conclusion logically flows from the premises because the premises are sufficiently definitive so as to create an inevitable conclusion.

The important point of this discussion is that we can use the deductive reasoning process to compel us to embrace conclusions that invariably result from the premises we create. But one can create statements that are deductive but that are inherently flawed. In a sense, then, the use of deductive reasoning is not completely symmetric because we cannot alter the premises with impunity and expect the logical consistency of the statement to hold true in all cases. Suppose that we offer the following sentence: "If you are a lumberjack, then you are a muscular man." This statement seems to make sense because we would expect the lumberjack occupation to attract muscular men. But we can reverse this sentence and obtain a completely different statement: "If you are a muscular man, then you are a lumberjack." Now even though there may be many lumberjack jobs available, not every muscular man chooses to go into the lumberjack profession. So the reverse of our original statement is not as persuasive and we are not compelled to reach the same conclusion.

This lack of symmetry in mathematical proofs becomes more apparent as we grapple with other "If, then" statements. After spending a few days riding a roller coaster, I would probably not be feeling very well but my experience would at least enable me to think about hypothetical statements such as the following: "If all Americans ride a roller coaster and I ride a roller coaster, then I am an American." Now this statement is structured in a way that seems to compel us to move from our premise regarding the preference of all American for riding roller coasters coupled with my own fondness for riding to the conclusion that I must therefore be an American. But there is nothing logically irresistible about this statement that requires us to accept its conclusion. The validity of this statement seems to be dependent on the assumption that my fondness for roller coasters will automatically cause me to be lumped into the class of Americans, all of whom like to ride roller coasters. But this conclusion is not logically inevitable because I do not have to be an American to ride a roller coaster. We would have to restructure our sentence somewhat in order for us to make the statement logically compelling, and we can do so in the following manner: "If Americans are the only people who like to ride roller coasters and I like to ride roller coasters, then I am an American." Although this sentence is a little more verbose than our previous sentence, we have set it up in such a way that the conclusion must follow from the premises. Our sentence is logically compelling because we have defined our premise (that Americans are the only persons who like to ride roller coasters) and a more specific statement regarding my own love for roller coasters to then allow us to move to the inevitable conclusion that I am an American.

Mathematicians like to approach all logical statements from the ground up. Here it is useful to think of the premises as the "bricks and mortar" with which these intellectual statements can be built. As a result, they do not use conclusory statements as a source of proof for the premises because the conclusions do not buttress the statements in a factual sense. Mathematicians understand that the conclusions are not actual proof (evidence) but merely the outcome of the reasoning process. To return to our roller coaster example, we would be engaged in a process known as circular reasoning if we were to take the conclusion that I am an American because I like to ride roller coasters and use that conclusion at the beginning of the reasoning process.

Although those who prize brevity might think that it would be a wonderful timesaver to be able to skip through the premises of logical reasoning and go straight to the conclusion (e.g., I am an American because I like to ride roller coasters), this approach would not be very helpful to us. We would find our so-called reasoning process reduced to nothing more than a series of random statements based upon our conclusions about what would be reasonable to conclude about a given situation. But this conclusion would not be one derived from a logically consistent reasoning process.

So it is important to remember that when we utilize deductive reasoning to construct a logical statement, we must begin with certain assumptions and then construct an argument that allows us to draw conclusions from those assumptions. Now there is admittedly an element of arbitrariness in our most basic assumptions in mathematics because we do have to begin somewhere if we want to get anywhere. In other words, mathematics is ultimately an intellectual edifice that overlays the phenomena of the real world and so we have to begin with certain elementary premises such as the field axioms we discussed earlier in this chapter. Otherwise, we would be unable to make any progress because there is no single starting point in which those who study mathematics can begin their work. So even though we may find it a little unsettling that we must, in a sense, build mathematics on "air," once we have made those initial assumptions we are able to carry out a surprisingly vast array of calculations and operations.

Most mathematicians like to simplify their equations wherever possible by cleaning up unnecessary clutter. One way to carry out a "clean-up" operation is to get rid of any unnecessary or redundant terms. A very simple example of this mathematical manicure may be illustrated if we wish to add the following equations: $2x + 2$ and $3x - 3$. Of course we could take the approach that "bigger is better" and merely string these two equations together into one bigger equation: $2x + 2 + 3x - 3$. But this bigger equation offers us very little in the way of insight; it is still nothing more than a collection or an aggregation of all the individual terms. This is not a satisfactory state of affairs and we must do something to simplify this equation. So we should combine like terms such as the two "x" terms ($2x$ and $3x$) and the two lone numbers (2 and 3) and thereby obtain the following solution: $2x + 2 + 3x - 3$, which can in turn be simplified as $5x - 1$. It is now neat and efficient.

A better example might be to take the binomial equation $(x + 1)(x - 1)$, which, when solved, is equal to $x^2 - x + x - 1$. We can see that the two middle terms of this solution ($-x$ and x) will cancel out. This will leave us with the comparatively sleek solution, $x^2 - 1$.

The importance of mathematical reasoning is less dependent on how the statements themselves relate to real-world objects than their own internal consistency. This is a fundamental feature of mathematics that we shall see demonstrated over and over again as we explore mathematics further. In order that we may strengthen our mathematical vocabulary, however, we should become acquainted with two terms: *lemma* and *corollary*. A lemma should not be confused with a lemming, which is a rodentlike animal. A lemma, by contrast, is a theorem that is used to simplify and thereby facilitate the proof of a subsequent theorem. We should recall that a theorem is a statement in mathematics to be proved. A corollary is a theorem that may be derived from a previous theorem. So mathematicians find that they use both lemmas and corollaries in constructing rigorous mathematical proofs because of the need to be efficient and to avoid reinventing the wheel already used by hundreds of their predecessors through the years. In this way, they can continue to build upon the formidable foundations of mathematics and, perhaps, create their own legacies of achievement and progress.

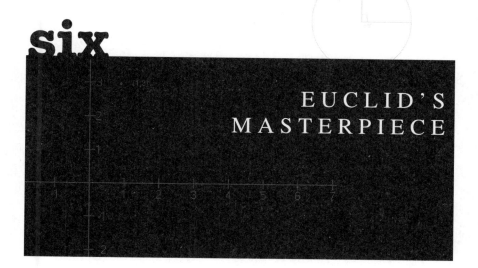

six

EUCLID'S MASTERPIECE

Any discussion of the story of mathematics must invariably include an account of the life and work of the Greek mathematician and historian Euclid. He lived in Athens for much of his life but moved to Alexandria in Egypt in the third century B.C.E. to head the mathematics department at the university established by the Egyptian king Ptolemy.[1] Euclid had gained great fame throughout the ancient world for his mathematical work and was one of the pre-eminent thinkers of his time. But he was also a very exacting scholar and was quite adept at the type of detailed work that would be so critical to his own literary labors. He is credited with having created the first geometry textbook—the *Elements*. Indeed, Euclid was the master synthesizer of the great mathematical works of many other mathematicians who lived throughout the Mediterranean World; as a result of this compilation, is often thought, erroneously, to be the creator of Greek geometry. According to British scholar W. W. Rouse Ball, however, the pedigree of his book is much more varied: "[T]he substance of books I and II (except perhaps the treatement of parallels) is probably due to Pythagoras; that of book III to Hippocrates; that of book V to Eudoxus; and the bulk of books IV, VI, XI, and XII to the later Pythagorean or Athenian schools."[2] Yet Euclid, as Ball acknowledges, did much more than to simply collect snippets of writings from notable mathematicians: "This material was rearranged, obvious deductions were omitted [and] in some cases new proofs substituted. Book X, which deals with irrational magnitudes, may be founded on the lost book of Theaetetus; but probably much of it is original [because] while Euclid arranged the propositions of Eudoxus he completed many of those of Theaetetus."[3] But even though it is not always clear where Euclid's summaries of other individuals works end and his own contributions to geometry begin, Euclid gave a valuable gift to the world because he provided a record of the mathematics of the ancient world that has survived to the present day. Of perhaps even greater importance is that Euclid arranged what had been a somewhat disjointed and jumbled subject into a pristinely logical presentation that has continued to profoundly shape the way in which geometry—indeed, all mathematics—has been taught to the present day.

Why is Euclid so important to geometry if all he did was gather together snippets of writings from itinerant scholars and mathematicians? Euclid not only preserved the legacies of earlier schools of thought but he also created an entire methodology for organizing geometry. His supreme accomplishment was to offer series of axioms from which much of the geometry of the ancient world could be deduced. As such, Euclid managed to devise the fundamental principles of geometry, which—although they had existed in various forms—had never really been distilled in such a straightforward and comprehensive manner. Consequently, Euclid could justifiably claim to have shaped and molded the subject of geometry and, indeed, be considered the creator of classical geometry. But the work is not without its criticisms: "The definitions and axioms contain many assumptions which are not obvious [and] no explanation is given as to the reason why the proofs take the form in which they are given [and] there is no attempt to generalize the results [and, among other things] the classification is imperfect."[4] But these criticisms, along with the traditional complaint that Euclid's *Elements* was unnecessarily verbose, should not obscure the fact that Euclid did create a unique form for the presentation of his propositions, "consisting of enunciation, statement, construction, proof, and conclusion."[5] Euclid was also responsible for "the synthetical character of the work, each proof being written out as a logically correct rain of reasoning but without any clue as to the method by which it was obtained."[6]

Perhaps the most striking thing about the *Elements* is that Euclid was not very concerned with the physical manifestations of geometric objects (e.g., circles, squares, spheres, cones). Indeed, a reading of the *Elements* would suggest that Euclid gave no consideration to the details of the real world. When he wrote about such things as points, lines, planes, and space, he was mainly interested in discussing the concepts themselves. Euclid would not have been very interested in the extent to which a string could be stretched taut to approximate a straight line, or a stone wheel could be smoothed like a circle, or a pyramid could resemble an equilateral triangle from a distance. "There is in Euclid the contempt for practical utility which had been inculcated by Plato [but] this contempt for practice was, however, pragmatically justified. No one, in Greek times, supposed that conic sections had any utility; at least, in the 17th century, Galileo discovered that projectiles move in parabolas, and Kepler discovered that planets move in ellipses."[7] Like Plato, who had been consumed with his distinctions between perceptions and abstract concepts, Euclid was more interested in theoretical principles than applications in the real world. No doubt he would have sympathized with Plato's preference for the pristine, immutable shapes of geometry (e.g., circle, square). But he was not greatly concerned about whether mere mortals would be able to recreate the physical counterparts to classical geometry's crown jewels.

What do you have to do to create a textbook that lasts for two thousand years? It helps if you have something original and profound to say to the world. In Euclid's case, his enduring contribution was to define the basic terms of geometry (e.g., points, lines, planes) to create a logical, self-consistent system of thought. But there were inherent limitations in such an approach because Euclid could not define each of his concepts in terms of ever-more basic concepts. He had to start with some arbitrary definitions so that he could avoid an infinite digression into an ever-greater etymological morass. In other words, he had to start by making certain arbitrary assumptions that would provide something of an intellectual bedrock for his subject. But Euclid was not always able to bring shimmering clarity to a concept as shown by his definition of a point as "that which has no part."[8] Needless to say, this definition did not inspire the poets to sing Euclid's praises, perhaps because elementary geometry has never been a popular subject for the great poets to plumb. But Euclid's definition was not greatly improved upon even by later writers such as Heron who, according to the mathematician David E. Smith in *History of Mathematics*, added the language "or a limit without dimension or a limit of a line."[9] Nowadays, we are more inclined to think of a point as something that is without length, width or depth—or having no dimensions at all. An illustrative example would be for you to imagine a pencil with an infinitely fine point that could be used to mark a dimensionless point. But it is virtually impossible for those of us who live in a three-dimensional world to think of a point in its purest geometric form because we must constantly remind ourselves that it is not a very small point but instead an infinitely small point—a subtle yet critical distinction.

Euclid then turned to the concept of the line, which he defined as "length without breadth."[10] This definition would have appalled Aristotle because it seemed to amount to a nondefinition that did not really help matters. However, Aristotle did not clarify the matter either when he defined a line as being a magnitude that could be divisible in one way. By contrast, a surface—in Aristotle's opinion—could be divisible in two ways and a solid divisible in three ways. The obvious point is that Aristotle was thinking of geometrical concepts in terms of spatial dimension—a line has one dimension, a surface (or plane) has two dimensions, and a sphere has three dimensions. Aristotle did not offer much in the way of clarification about elementary geometry. But he did favor a sort of geometrical hierarchy whereby a succession of points arranged in a linear fashion would produce a line; an array of lines would create a plane (surface); and, by implication, a surface in motion would give rise to a solid. It is this sort of ranking that later figured in Euclid's work because one had to begin with the point and then use points to construct lines and lines in turn to construct planes and so forth. But Euclid's treatment of the subject matter was far more systematic and detailed than that offered by Aristotle.[11]

Once Euclid began to provide some flesh for these basic geometrical definitions, he felt free to use them in various ways in order to create more profound geometrical concepts. He realized that it did little good to offer up unclear definitions of points, lines, surfaces, and solids if one did nothing further with them. So Euclid's first order of business was to express the concept of the "straight" line as "a line which lies evenly with the points on itself."[12] This definition was less a clarification than an obscuration. Other commentators such as Archimedes would later try to shift the scope of the analysis by mashing the concepts of "line" and "distance" together to say that a straight line is the shortest distance between two points. And while this definition would suffice for nearly twenty centuries, it, too, would be called into question when nineteenth-century mathematicians such as Nikolas Lobachewsky and William Kingdon Clifford devised negatively and positively curved spatial geometries in which it was impossible for a straight line to be the shortest distance between two points.

Given the ease with which the ancients fumbled around with the most basic geometrical concepts, it is not surprising that Euclid and his brethren were not always successful in grappling with two- and three-dimensional concepts such as surfaces (planes) and solids. In general, Euclid's predecessors had a notion of the surface concept that was probably not unlike their notion of pornography— that is, they (like the United States Supreme Court) could not define it but they knew it when they saw it. Perhaps the most successful metaphor was offered by the Pythagoreans who decided that a surface was most like a "skin" or "color." Certainly it is easy for mathematics students to think of a surface as a skin because it is like a sheet of paper. It has both length and width but virtually no depth. For his part, Euclid limited himself to defining a surface as "that which contains only length and width."[13]

The surface concept was further refined as the ancients tried to understand the idea of a plane surface—one that extends indefinitely far in all directions like the surface of the ocean, but lacks depth. Euclid himself suggested that "a plane surface is [one] which lies evenly with the straight lines on itself."[14] No doubt Euclid was trying to underscore the perfect flatness of such a geometrical abstraction. He seemed to be suggesting that a plane surface could only consist of a collection of straight lines all lying flat next to each other, rather like a row of nails in a box. But he found himself hard-pressed to express the concept of a plane surface in a way that did not lapse into confusing references to extended and continuous lines. However, he may have found Aristotle's definition of a surface as a line in motion to be of some assistance.[15]

Of course language was the linchpin of Euclid's masterpiece. When you try to present a subject such as geometry for the first time, you must offer a panoply of defined terms (e.g., point, line, plane) and thereby create a new language for understanding exactly the concepts and ideas you are trying to convey to your readers. Conversely, the reader must plunge into the textbook without fear and try to master the language. "The propositions of Euclid are arranged so as to form a chain of geometrical reasoning, proceeding from certain almost obvious assumptions by easy steps to results of considerable complexity."[16] Any progress requires that the reader master the terminology. Otherwise, any attempt by the reader to conquer the subject will run aground because he or she will lack the requisite "language skills" to move from the most basic concepts to the more advanced ideas. Although the *Elements* was often criticized for the laborious detail that it lavished on often seemingly simple concepts, it did provide a sort of intellectual "toughening" process for the reader desiring to master geometry. The fact that it lasted for almost two thousand years as *the* textbook on the subject is perhaps the most indisputable evidence of its worth.

Much of Euclid's work is arguably static. As we have seen, he painstakingly (though not always clearly) defines concepts such as points, lines, planes, and surfaces, and offers detailed geometric proofs. This is all very nice and informative but it is somewhat ossified. There is little drama associated with talking about points and lines. Fortunately, things pick up a bit when Euclid begins to deal with more sophisticated geometrical concepts. The first arguably dynamic element of Euclid's geometry is his effort to define an angle. He views an angle as being created by the convergence of two nonparallel lines with the angle being represented by the amount of inclination between the two lines. The amount of declination may be better illustrated if you imagine yourself to be a glider pilot flying above the terrain. Although we know that the earth is round, we can usually consider it to be a flat plane for short distances. If we are flying a level flight, then our flight path is essentially parallel to the surface of the earth; our plane will neither come closer to the surface nor move farther away from it. So we can enjoy the quiet silence of our panoramic view of the world, knowing that we are flying perfectly parallel to the planet. But we know that such a flight path is not a good plan if we wish to demonstrate the concept of the angle. Knowing full well that we must press forward in order to advance the frontiers of human knowledge, we might don a crash helmet, say a silent prayer, and push the throttle forward. Our glider would then nose downward and hurtle toward earth. We could then hit the ejector seat and float slowly to earth as we watched our glider spiral downward. If the glider plunged into the earth and plowed along the ground, then the angle or "slant" at which the glider hit the ground would be the angle of inclination. If our glider had plunged to earth straight down, then its angle of impact would be a right angle because its downward path would be perpendicular to the surface of the earth. Of course most pilots bailing out of crashing planes are not overly concerned whether the plane hits the earth at a right angle or some other angle. But we can agree that the angle at which our glider hits the earth will represent the amount of "inclination" between its surface and the path of the glider's descent.

Now there is something to be said for learning a little bit more about angles and the terminology associated with these angles. But all knowledge about angles begins with learning that a circle consists of 360 degrees and that all angles are, by definition, less than 360 degrees. Knowledge about the geometry of the circle is one of the most profound secrets of the ancients. But there are indeed even more delights awaiting you as we move throughout many areas of the mathematical realm. But we need to think about this idea of a circle having 360 degrees because the measure of degrees is essentially a measure of movement along the circumference of the circle from a single starting point. If, for example, we travel from a point on the circle a distance equal to ? of the way around the circumference of the circle, then we will have traveled ? of 360 degrees or 90 degrees. If we then imagine digging two holes, one from the original starting point of our journey to the center of the circle and the other from our new location 90 degrees away down to the center of the circle, then the convergence of these two holes will form a right angle of 90 degrees (see figure 1).

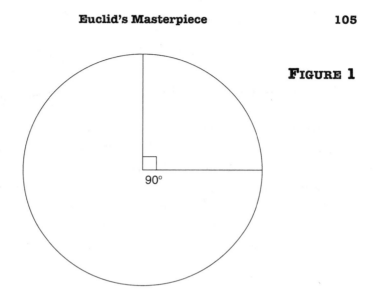

FIGURE 1

What if our journey from our original starting point is much shorter? Suppose that we set out on our journey but had traveled only ⅟₂₀ of the circumference of the circle or 18 degrees by the time we reached our destination. If we dig down to the center from this stopping point, then we would intersect the original tunnel at an angle known as an acute angle, which would measure 18 degrees (see figure 2).

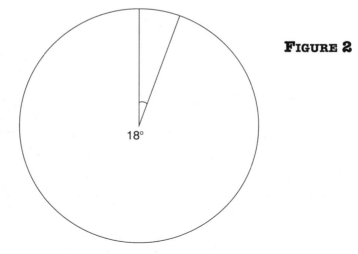

FIGURE 2

It remains for us to consider the outcome if we had decided not only go beyond the 90 degree mark (¼) of the circumference of the circle, but indeed continued even farther, passing the 180 degree mark (½) of the circumference before stopping to rest. Upon taking a measure of our final stopping point, we might find that we had not only passed the halfway point of the circumference but had made it exactly ¾ of the way around the circumference. This would represent something of an accomplishment since we would find, upon digging our tunnel to the center of the circle, that the intersect of our tunnel with the original tunnel would give us an angle of 270 degrees (see figure 3).

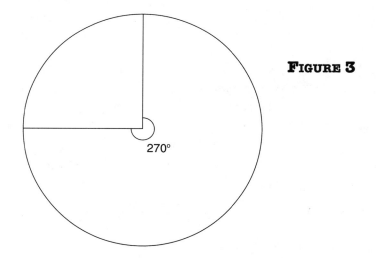

FIGURE 3

This angle is known as an obtuse angle. The reference to the word "obtuse" does not refer to the difficulty of understanding this concept. Instead it is easier to think of it as an angle that is not acute because its measure of inclination is comparatively large. It is not sleek and svelte like the acute angle. But it is no more or less extraordinary than the acute angle because both are dependent on the amount of movement from a fixed point along the circumference of a circle.

More recently, mathematicians have viewed angles as far more dynamic entities than Euclid and his colleagues would have ever imagined. Instead of simply considering angles as abstract measures in which one line intersects with another, the more modern view is to think of an angle as a measure of turning. Instead of going to all the trouble to travel from a point of origin to a certain destination along the circumference of the circle, more modern thinkers conserve their strength and merely turn their bodies so that their field of vision moves from the imaginary line at the point of origin to the point of destination. I can better understand this idea of turning by using my own body in the following manner: Suppose I am facing due north when I begin this experiment. This is my point of origin. If I then turn my body until it faces due east, then I have turned ¼ of a circle, or 90 degrees.

Certainly this focus on turning as opposed to walking along the circumference of a huge circle offers a more easily understood concept of the angle. If we make one-half turn when starting at our point of origin, then we will be facing a point that is exactly 180 degrees apart from the starting point. If we lose interest in this exercise and only turn ¹/₁₀ of a complete circle from the point of origin, then we will find ourselves staring at a point exactly 36 degrees away from the original point of origin. Needless to say, this exercise in turning can be repeated thousands of times with any fraction of the circumference of the circle.

We could move on to the actual definition of circles offered by Euclid and his cohorts, but you might find the definition, though illuminating, to be still lacking. Euclid defined the circle as a figure in a single-surface plane in which a point designated as the center of the circle contained all straight lines of equal length lying within that plane. As the lines all radiate outward from that single point in all directions and are of equal length, then they must necessarily form a circle. Despite the passage of twenty-three centuries or so, our finest mathematical minds have not really been able to devise a better definition of the circle. And so it continues to be defined in a

rather murky way to the present day. But we can live with the definition of the circle offered by the ancients because, quite frankly, we
probably would not be able to do a better job. Moreover, modern
mathematicians have basically agreed to accept the limitations
inherent in the definitions of most basic geometric concepts. After
all, much of their own work is built upon the foundation provided by
Euclid and other ancient mathematicians and, like it or not, they
have no interest in undermining this edifice. Certainly there is no
widespread demand for rebuilding mathematics from scratch.

Given the lack of clarity regarding the circle itself, it is not surprising to see that the terms "diameter" and "radius" have been muti-
lated by historians as well. Euclid defines the diameter of a circle as
a line bisecting the circle. This bisecting line splits the circle into two
equal halves. Euclid was evidently very pleased with his definition of
diameter, so much so that he did not really offer a definition for
radius. He tossed out the term "distance" to mean "radius," but it was
obviously not something that was uppermost among his concerns.
Indeed, David Smith suggests in his *History of Mathematics* that the
earliest recorded use of the word "radius" occurs as late in the sixteenth-century in Ramus's *Scholarvm Mathematicarvm, Libri vnvs et
triginta*.[17] Other scholars such as Boethius, while not referring to the
term "radius" specifically, did use language that appeared to embrace
the concept of a "semidiameter."[18]

We conclude this chapter with a brief look at Euclid's definition
of parallel lines, which is of such paramount importance to classical
geometry. Euclid offered a definition of parallel lines as straight
lines that lie in the same plane and extend indefinitely (forever) in
both directions but still do not meet one another in either direction.
If you think about it, the definition makes sense. Suppose that you
and your neighbor are each holding an infinitely long pole and are
standing next to each other. If the two of you are holding the poles
at the same height, then it becomes very clear that the slightest turn
of one pole toward the other will cause the two to converge at some
point in the distance. And if you try to point that end of the pole

away from your neighbor, then you will be unpleasantly surprised to find that the part of the pole extending behind you will at some point cross the pole of your neighbor. As with many of Euclid's definitions, later commentators have not greatly improved upon Euclid's prose and so we are best left with relying on his profound words.

Although Euclid's definitions permeate the geometry textbooks of the world's students, they represent only a part of Euclid's contribution to mathematics. Indeed, an arguably more valuable feature of his *Elements* was his offering of a number of principles to guide students of geometry. These so-called axioms and postulates provide the skeletal structure of geometry as well as the basic assumptions underlying all geometry making it possible to construct an entire mathematics based upon logically consistent statements. But they would not be very useful if Euclid and other early mathematicians had not used logical reasoning to create a mathematical framework that would last forever. Both this reasoning process and the principles guiding the creation of classical geometry shall be explored in the following chapter.

seven

GEOMETRY AND MATHEMATICAL REASONING

Our previous dealings with Euclid revealed him to be a fine synthesizer and expositor of the basic concepts of classical geometry. But definitions by themselves have limited use because they do not help the mathematician to do anything other than draw shapes and make certain limited conclusions about the relationships between these concepts. But something more in the nature of a superstructure is needed to breathe life into what would otherwise be little more than an exercise in taxonomy. Fortunately for those who worry about the integrity of mathematics, Euclid was far ahead of the game and took the additional step of devising a number of principles or axioms that could not be proven, but could provide a foundation upon which the basic ideas of geometry could be constructed. Euclid was not shy about his ambitions for systematizing geometry and so he ventured headfirst into the fray. He offered not only axioms that were fundamental truths applicable to all sciences but also postulates, which, though not designated as such by Euclid, were applicable to specific branches of the sciences.

Euclid's axioms were general statements or truths that were not, at first or even second glance, particularly relevant to geometry. Indeed, they were general propositions that might have well been applicable to biology or physics but Euclid saw fit to begin with these very same axioms in his treatise on geometry: (1) all things that are equal to the same thing are equal to each other; (2) all things that are equal and are added to other equal things will yield wholes that are equal to each other; (3) when equals are subtracted from equals, the remainders are equal; (4) all things that coincide with one another are equal to one another; and (5) the whole is greater than the part.[1] These axioms could have just as easily appeared in a martial arts movie or in a garden club presentation because they really do not relate to anything in particular. Euclid, to his credit, relied on a largely self-consistent set of axioms in constructing his masterwork, which, given the near universality with which it was accepted for two millennia, is quite remarkable.[2] Certainly it was in stark contrast to many fields in which the basic principals underlying the body of knowledge have been superseded, discarded, and revised over and over again.

Why did Euclid make these arbitrary assumptions? It was as much a matter of practicality as anything because he has to start somewhere to construct a system of thought. To avoid simply wasting an enormous amount of time, the easier route is to adopt certain statements so as to create a sort of "ground-floor" set of presumptions and then build the edifice upward from that point. Euclid was also aware that there was a large number of arguably superfluous axioms being bandied about the ancient world by individual mathematicians. Euclid himself seems to have felt that logic demanded that all of geometry be constructed with as few a number of axioms as possible. After all, there is a beauty in finding the basic constituents of a subject. Moreover, there is a certain satisfaction to be gained from the step-by-step construction of a mathematical concept. So Euclid's search for axioms was driven by the desire to assemble a set of basic principals about which all reasonable mathematicians (even the Pythagoreans) could agree were essential to the creation of geometry.

The general scope of coverage of Euclid's textbook then shifted to the establishment of certain theorems derived from purely deductive reasoning. Indeed, all of mathematics is based upon deductive reasoning, so Euclid was not being a wild-eyed revolutionary in using the deductive process. Unlike inductive reasoning which involves gathering information about the general state of the world to draw specific conclusions, deductive reasoning, as noted earlier, is based upon the formulation of certain specific premises from which general conclusions can be drawn. In essence, it is the reverse of inductive reasoning. An example of deductive reasoning might involve the following premises and conclusion: "All dogs bark. There is a dog. Therefore, that dog barks." Of course you may be a "cat person" and therefore unable to appreciate the value of such an example, so you could substitute the word "cat" in place of "dog" and the word "meow" in place of "bark" to get the following logical argument: "All cats meow. There is a cat. Therefore, that cat meows." Because the premises of this argument are factually correct, the conclusion regarding the sound made by the cat is indisputable. But from a logical point of view, it would not make much difference whether we said that the cat meows or the cat barks because the structure of the argument is such that we have stated that all cats constitute a single class having the same characteristic. Furthermore, we have identified a single member from that same class. As a result, that member must necessarily have the same characteristic as all other members of the class because it is by definition a member of that class. The validity of our premises, by contrast, depends on the worth of our real-world observations because we cannot construct meaningful logical arguments if the premises are not rooted (at least to some degree) in reality. If we propose a nonsensical argument ("All cats bark. There is a cat. Therefore, that cat barks."), then we maintain the logical consistency of our statement but we create a statement that has no real value.

As we pointed out before, Euclid was very fond of deductive reasoning in his *Elements* and used it constantly in constructing hundreds of theorems. Although some students have viewed these theorems as a gigantic plot to toy with their minds, Euclid's motives were less sinister. Deductive reasoning served as a sort of construction crew whereby the edifice of Euclid's geometry could be erected. Regardless of how watertight his reasoning, however, Euclid had to start, as we have said, with certain basic assumptions from which his geometry could be formulated. Although geometry itself is conceptual and abstract, its groundings are, paradoxically, rooted in commonsense experiences. The assumptions from which we derive the basic principles of geometry are not ones that can necessarily be proven as we are really trying to formulate certain statements about which reasonable people can agree. The strange thing about these statements is that we do not necessarily care whether these statements are true, nor do we even bother trying to verify them. Admittedly, this is a sort of intellectual nihilism that may not inspire confidence among all readers of geometry textbooks but it is basically unavoidable. The old proverb that the longest journey begins with a single step could apply to geometry because these statements represent single steps. Let us consider these assumptions in greater detail.

AXIOMS AND POSTULATES

These assumptions are often divided into two groups—axioms and postulates. Axioms are general assumptions that govern geometry as well as much of the rest of the mathematical terrain, including arithmetic and algebra. This universality means that axioms are important things for all students to study because they are so pervasive and their importance must be appreciated to understand the evolution of mathematics itself. Most basic axioms are very simple, commonsensical statements. In other words, we do not require proof of the axiom that any quantity is equal to itself or that the whole of a quantity equals the sum of its parts because we think that these statements are both reasonable and acceptable at first glance. What we need to avoid is reading more into these axioms than can be reasonably justified.

When we look at the axiom that the halves of a quantity are equal, we need to simply accept it. There is nothing to be gained by wondering about the philosophical ramifications of two halves being equal to one whole, for example, because we have already arbitrarily defined one whole as consisting of two equal halves. Indeed, it is more helpful for mathematics students to view axioms as general principles that will provide some boundaries for our survey of mathematics. From Euclid's standpoint, these axioms were akin to a roadmap whereby the construction of geometrical principles would be both directed and constrained.

The axioms we discussed above are not very numerous and we do not require large numbers of such axioms because they do not really relate to the detailed operations of geometry. By their very general nature, they could be equally applicable to almost any other branch of mathematics. Postulates, by contrast, are general assumptions that are specific to geometry itself. We should point out, as noted by the American mathematician Carl B. Boyer, that "modern mathematicians see no essential difference between an axiom and a postulate."[3] So when a mathematics student hears the statement that "only one straight line may be drawn through any two points" or that "two distinct straight lines cannot have more than a single point in common," he will know that he is hearing a postulate. All you need to remember is that *axioms* are general statements about (virtually) *all* of mathematics. In contrast, *postulates* are statements about only a *particular* area of mathematics (geometry).

Many of the most basic postulates will seem vaguely familiar to those persons who were fortunate enough to learn about Euclidean geometry in high school. These postulates should seem like long forgotten friends because they are essential for geometry. They tell us that a line can be extended to any distance and that all diameters of the same circle are equal. Other postulates tell us that the shortest distance between two points is a straight line and that only one straight line may be drawn that is parallel to another straight line. Of course this book is not intended to merely spew out the postulates of every branch of mathematics. Moreover, it would be beyond the scope of this book to cover every postulate in detail. Our most important point is that geometry is a creature of these postulates and that they provide the foundation upon which the hundreds of theorems of Euclidean geometry can be proven.

The actual number of postulates in Euclid's *Elements*, like the number of axioms described above, are few in number and are as follows: "(1) to draw a straight line from any point to any point; (2) to produce a finite straight line continuously in a straight line; (3) to describe a circle with any center and radius; (4) that all right angles are equal; and (5) that if a straight line falling on two straight lines makes the interior angles on the same side less than two right angles, the two straight lines, if produced indefinitely, meet on that side on which the angles are less than the two right angles."[4] These postulates were evidently driven by Aristotle's belief that the superior mathematical proof would be the one with the fewer number of postulates and Euclid was apparently one who embraced Aristotle's preference for the simple.

But these postulates seem to amount to more of a series of objectives or even a sort of mathematical philosophy. Indeed, one is struck by the general nature of these postulates, which immediately raises the question as to how one is supposed to do the detail work in mathematics using such principles. After all, mathematicians cannot be expected to accept everything on faith. Theorems differ from both axioms and postulates in that they are formal propositions that must be proven. The proof itself may begin with an axiom or a postulate or a definition or a hypothesis and then proceed through a series of steps, often involving additional axioms or postulates, to a conclusion. The proof may be accompanied by a diagram to identify a particular line or angle. A well-known example of a mathematical proof is the one used to prove that two angles of an isosceles triangle are themselves equal. An isosceles triangle is a triangle that has two longer equal lines that join together at a point on top and are connected on the bottom by a shorter single line (see figure 1). This particular proof offers us a good example of the way in which mathematicians approach problems, using deductive reasoning to move from one step to the next. Despite any misgivings we may have about treading into formalistic terrain, we should try to review this proof so that we, too, can gain some insight into the methodology used by mathematicians.

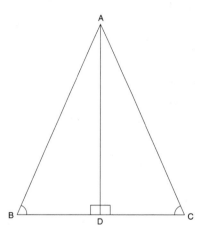

FIGURE 1

Our triangle has two equal sides, AB and AC, as shown in figure 1. We want to prove that angle C of the triangle is equal to angle B. Our effort to construct this proof requires us to draw a line AD that will cut triangle ABC in half, with D being the midway point between B and C. This line will split the triangle into two equal halves ABD and ADC, which we know have equal areas because our hypothesis states that lines AB and AC are equivalent. We also know that the two sides of angle A in BAD and CAD are equal because of the axiom that provides that the two halves of a quantity are equal. The fact that the two sides and the angle of BAD and

CAD are equal is underscored by the postulate that holds that two such triangles are congruent (equal) to each other. This enables us to determine that angles B and C are equivalent to each other because another postulate tells us that corresponding angles of congruent triangles are themselves equal.

Please bear in mind that mathematical proofs are of paramount importance because they illustrate the way in which problems are solved in a logical manner. Indeed, this form of reasoning is arguably of fundamental importance to all of Western civilization. This statement is not so far-fetched a statement as it may first appear when we recall that almost every branch of knowledge—ranging from physics and astronomy to mathematics and chemistry—is dependent on the integrity of the deductive reasoning process. We are able to employ deductive reasoning because it offers a reliable means of intellectual inquiry that progresses from premise to conclusion based upon a consistent line of thought. If we were unable to use deductive reasoning, then much of our knowledge would be rendered useless or at least much more difficult to use in a profitable manner. We would have to rely on more cumbersome methods for drawing conclusions from our hypotheses, such as trying to prove each step empirically in a proof instead of relying on the consistency of the steps themselves.

For mathematicians and historians, geometry is ultimately based upon the viability of mathematical proofs that are themselves dependent upon logical analysis. This might seem a little short-sighted because this view limits mathematicians to deductive reasoning alone and necessary precludes both inductive reasoning and the more haphazard method of reasoning by analogy. But there is a compelling justification for this restriction of mathematical theory to deductive reasoning that becomes clear as we consider the nature of both inductive reasoning and reasoning by analogy.

REASONING IN MATHEMATICS

The person who relies on inductive reasoning looks around to try to uncover recurring patterns or events in the subject matter that she is trying to investigate. Suppose, for example, that Aristotle is out for a walk and happens to spot an egg in a bird's nest. Upon closer examination, he discovers that a baby bird is slowly pecking through the egg. Because Aristotle is a first-rate philosopher who is very familiar with deductive reasoning, Aristotle might conclude that all birds come from eggs. However, one cannot draw a conclusion from only one observation using inductive reasoning. Aristotle would have to head back into the woods and search for other birds' nests to see if birds hatch from eggs. We would expect that he would find that his initial suspicions about eggs as the source of baby birds to be confirmed. In this way, Aristotle would become convinced of the truth of his conclusion that birds hatch from eggs as he peered into nest after nest. So Aristotle's general conclusion that all birds hatch from eggs would be based upon his repeated observations of hatchlings pecking their way out of their eggs. This would be a sort of "weight of the evidence" approach in that Aristotle would give greater credence to his hypothesis about birds and eggs as he made more and more observations. However, the problem with inductive reasoning is that Aristotle would never be able to be completely sure as to the accuracy of his prediction until he had observed every single bird hatching from every single egg. Thus there is a vaguely unsatisfactory aspect to inductive reasoning due to the inherent impossibility of verifying the truth of all possible observations.

But what about reasoning by analogy? If Aristotle had been laying around watching birds hatch from eggs, he might have concluded that all other animals hatch from eggs. Aristotle then might argue that the fact that birds hatch from eggs should provide a model for other deserving animals. But as you can imagine, reasoning by analogy is a tricky business because it is based on what we think at first glance should be a reasonable state of affairs for other situations. One can imagine Aristotle wandering into a lion's den in search of lion eggs and meeting his untimely end. In any event, the point of this digression is that it is not a precise science by any means to reason by analogy because it necessarily compels you to focus on the similarities of two situations and, accordingly, ignore the differences that may prove to be insurmountable.

Mathematics itself is basically a product of deductive reasoning. Every mathematical proof consists of a series of deductive arguments and every deductive argument in turn consists of premises and conclusions. It is the integrity of this reasoning process that forms the intellectual "spine" of all mathematics because it does not, in its pure form, purport to relate to or draw upon real world examples. By this statement, we mean that the deductive approach does not require one to look for events or phenomena in the real world and then draw conclusions as would occur through inductive reasoning. It instead relies on our making certain assumptions that are strung together in some sort of coherent fashion so that certain conclusions can be drawn. But the reliance of mathematics upon deductive reasoning provides it with a certain irresistible quality because our acceptance of the premises of a mathematical argument necessarily compels us to accept the conclusion that arises from those premises.

In its most basic form, deductive reasoning, as we have seen previously, can be structured in the form of an "If, then" type of argument: "If Matt is a politician and all politicians are vegetarians, then Matt is a vegetarian." If we found that Matt liked to eat only raw meat and had been tossed out of the local vegetarian club, then we would have to question the validity of the basic premises. If the premises were incorrect, then the conclusion would necessarily be incorrect. But if we found that Matt had decided to renounce eating raw meat and was still pursuing his political career, then our premises would be correct and our conclusion about Matt's preference for a vegetarian diet would necessarily follow.

Yet not all statements are necessarily structured in a way that compels us to accept the conclusion. We can have "If, then" statements set up in such a way that the premises do not necessarily lead to the conclusion: "If all boats float and my dog floats, then my dog is a boat." Even though this statement has the same "If, then" format as our previous example dealing with Matt the vegetarian, its conclusion does not invariably follow from the premises. The fact that my dog floats does not entitle me to call it a boat. But the problem with this conclusion stems from the premises themselves because dogs and boats are two entirely different things; the fact that all boats float and my dog floats does not necessarily mean that my dog is a boat.

The axioms and postulates of geometry as presented by Euclid are clearly very abstract in nature and do not seem to bear any tangible relationship to the physical world. Certainly Euclid would not have been interested in the practical applications of geometry, such as building houses or roads. But the advantage of this deductive approach is that it allows one to ignore the messiness of the physical world and to follow the natural progression of the logical argument itself until the very end. If the consistency of the reasoning is scrupulously followed, then the result is a certain outcome. But the fact that one is beginning with premises rooted in abstract principles (e.g., points having no size, lines of infinite length, perfectly round circles), as opposed to real-world observations, means that there is a detachment from the physical universe. Inductive reasoning and reasoning by analogy are admittedly less precise, even sloppy, when compared to deductive reasoning. Yet they have their own strengths in that they are rooted in the imperfections of the real world and are not limited to the matters that are subject to examination by deductive reasoning. Moreover, many branches of science, such as astronomy, chemistry, physics, and biology, are dependent on elements of inductive reasoning and reasoning by analogy for much of their advancements. Certainly the chemist who mixes different combinations of substances together to try to create a new and better form of latex, for example, may necessarily use inductive reasoning based on the results of past experiments gone bad and reasoning by analogy based upon previous efforts to create cutting-edge latex products. But these fields necessarily require that the scientist interact with his or her environment. After all, a chemist can do little good merely by drawing equations on a board that he believes will yield a new and improved latex substance. He must also conduct experiments to determine whether the process actually works.

Although deductive reasoning admittedly has its limitations, it does offer certainty for those who follow its logical structure to the end. The fact that its utility may be confined largely to areas of knowledge in which abstract principles generally transcend the mundane world of our existence should not cause us to complain. Indeed, it is the fact that humans can use reason that may be the most important distinction between ourselves and the animals of the world.

The Greeks, not surprisingly, were the first civilization to recognize the incredible power of deductive reasoning and the way in which it could extend far beyond the traditional limitations of the ordinary human senses of sight, sound, smell, taste, and touch. Whereas our ordinary sensory experiences were necessarily constrained by the scope of our experiences (what we see, hear, taste, touch, and smell), the power to reason altered humanity's perception of the universe forever. It would no longer be a world in which we responded solely based upon our own sensory interactions with our environment. The advent of reason changed us into beings who could ask those pesky "why" questions that have so bedeviled philosophers to the present days—questions such as "Why are we here?" "Why is there suffering in the world if there is a God?" "What lies beyond the universe?" and so on. Even though these questions really did not have answers that could be derived through any means of logical reasoning available to us, they did force us to look at the world around us and consider it as something more than just a stage in which we played out our lives and died. It forced us to consider the reasons we might have for being on this planet in the first place and to consider issues such as life and death and good and evil. The human mind could consider the infinite even though we would never actually see the infinite. The human mind could imagine numbers so large that one could fill the observable universe with their zeros even though we would seldom encounter groups of objects in our daily lives numbering more than a few hundred or thousand at a single time (unless we considered such things as the number of drops of water in the ocean or the number of leaves in a forest). But reason freed the mind from any physical constraints and made it possible for us to imagine grandiose visions of the cosmos dwarfing anything that has ever been observed. Some people believe, for example, that our universe is but one of an infinity of such universes even though we would be hard-pressed to imagine how an infinity of infinitely large galactic systems would exist together in space. The fact that we cannot imagine how such a system would exist does not necessarily preclude us from imagining that it might exist.

Geometry provides a tapestry in which the human mind is free to roam and create an entire abstract universe. Although geometry did not begin with the Greeks, it is certainly true that our modern study of geometry bears a distinctly Greek imprint. The geometer begins with comparatively few axioms and postulates—much like the artist who is able to create countless variations of colors and shades with just a few colors and instruments—and constructs deductive arguments that can be linked together to form many different theorems. These theorems in turn would become part of the knowledge base available to geometers who could then use them to construct additional deductive arguments, virtually without limit. But the important point is that this approach enabled the Greeks to develop a theory of numbers and not simply utilize mathematics for its purely mechanical applications such as counting sheep or building pyramids. Although the Greeks may have relied too heavily on deductive reasoning to the exclusion of real-world considerations, they did elevate the study of mathematics to an unprecedented level of sophistication that certainly would have been inconceivable in those earlier societies in the ancient world whose only use for mathematics was for the solution of purely practical problems.

eight

ONE, TWO, THREE, TRIGONOMETRY

Trigonometry is a field of mathematics that occupies a sort of curious no-man's land in its prominence in the education of most students despite its numerous practical applications, most notably in astronomy and geography. The subject is usually taught to high school students after they have completed the three years of mathematics that is required by most high schools for graduation. Unfortunately, most students do not venture far enough in their mathematical studies to learn about trigonometry—even though they might find trigonometry to be far more accessible and useful to them than they would have ever believed possible.

Trigonometry provides the user with the intellectual tools to do many useful things, such as determining the height of a mountain, the distance to the moon, or the dimensions of a piece of land. In its most fundamental sense, trigonometry is the study of triangles. It literally means to measure a triangle and its practictioners study the relationships between the lengths of the sides and the measures of the angles of triangles. Although many people can remember rolling their eyes when they were told by their teachers that mathematics would be of great help to them in their careers, it is actually quite true that trigonometry is a prerequisite for many careers including engineering, surveying, and architecture. Also of importance is the fact that an immersion in trigonometry strengthens one's knowledge of geometry and helps one to become more adept at discussing quantitative matters and abstract concepts.

Triangles are the be-all and end-all of trigonometry. But you are no doubt wondering, What is so special about triangles? After all, they are very easy to draw and they seem so simple. But the triangle can seem deceptively complex when analyzed from a mathematical standpoint. To appreciate the basic principles of trigonometry requires that we understand the structures of triangles themselves.

First, we need to point out that a triangle has three sides. Each of these sides intersect with the other two sides, meeting at points which are called vertices. Each vertex (the singular form of vertices) forms a point of the triangle. Not surprisingly, each vertex has an angle. In considering triangles, mathematicians have traditionally used upper-case letters (A, B, C) to designate the vertices, and lower-case letters (a, b, c) to designate the lengths of the triangle. This use of alphabetic symbols is not intended to terrify students. Instead it is a convenient way to describe the particular part of the triangle you wish to examine without having to worry about extraneous commands such as "the line to the left of that point on the top—no, no, *that* point." It enables us to be precise and sparse in our use of language and to know what we are talking about without having to explain it in great detail.

What we now call trigonometry originated with the work of the Greek mathematician Hipparchus in the second century B.C.E. "The basis of Hipparchus' astute method is a simple theorem of geometry [that states] that if two triangles are similar, the ratio of the lengths of any two sides of one triangle equal the corresponding ratio of the other. Thus if triangles ABC and A'B'C' are similar, for example, then BC/AB equals B'C'/A'B'. If triangles ABC and

A'B'C' are right triangles and if angle A equals angle A', then . . . we know that the triangles are similar. [Therefore] the ratio of the side opposite angle A to the hypotenuse of the triangle must be the same for any right triangle containing angle A."[1] What we know as the study of trigonometry involves the calculations of various ratios that can be formed in a given triangle based upon the arrangement of different sides of the triangle which are given, as we shall see, names such as sine, cosine, and tangent.[2] Having knowledge about the properties of a triangle makes it possible for mathematicians to use trigonometry to calculate distances and heights that could never be measured by hand as demonstrated below.

When wading into the waters of trigonometry, we should begin with the right triangle—which is probably the triangle most familiar to us. It is also perhaps the most helpful triangle in the practical applications of trigonometry. It is called the "right" triangle because it boasts a right angle of 90 degrees. Angle A in figure 1 is a right angle because one line intersects perpendicularly with another at a 90 degree angle. For example, the walls in a room that has four equal sides typically intersect with each other at 90 degree angles.

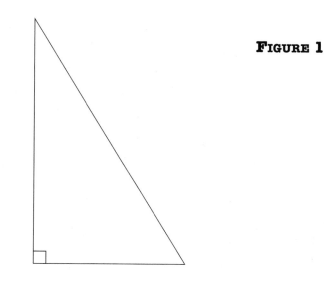

Figure 1

Now it is important to recall that the angles of every triangle—no matter how it is shaped—have a total of 180 degrees. You will immediately recognize that this fact means that if one angle of a right triangle is equal to 90 degrees, then the other two angles added together must also be equal to 90 degrees. Of course there are some people who will wonder if you can have a triangle with two right angles. The only problem with that approach is that it does not allow for the third angle to have any degrees at all. In short, you end up with a geometric figure that cannot be drawn. So we have to cast aside the search for that rare two–right angle triangle and instead turn back to figure 1.

The wonderful thing about right triangles is that the other two angles will necessarily be acute angles because of the need for them to add up to 90 degrees. But it does not follow that these two angles will each be equal to each other. Why not? The measure of the angles necessarily depends on the length of the legs of the triangle. If the two legs that intersect to form the right angle of the triangle are equal in length to each other, then we will have what is called a right isosceles triangle. The common lengths of the two legs means that it will have three angles that measure 90 degrees, 45 degrees, and 45 degrees, respectively. An equilateral triangle, by contrast, is one in which all three sides are equal in length and, hence, all three angles are equal—each being equal to 60 degrees.

How do we determine the lengths of the various sides? First, we need to recall the Pythagorean theorem; generally credited to our Greek friend Pythagoras—which states that the sum of the square of the legs of any right triangle is equal to the sum of the length of the hypotenuse. In other words, side a^2 + side b^2 = side c^2. So if we have a triangle with legs equal to 3 feet and 4 feet, respectively, then the Pythagorean theorem tells us that $3^2 + 4^2 = 25 = 5^5$. We are thus able to determine that the hypotenuse is equal to 5 feet. The outcome will be a little different with a right isosceles triangle because the lengths of the legs of the right isosceles triangle always have the ratio 1:1:$\sqrt{2}$. So if the two legs are each equal to 1 inch, the third side (the hypotenuse) will be equal to $\sqrt{2}$ inch—the square root of 2, which, as any mathematics student knows, is equal to 1.414213562. Any time we encounter a problem in which we know the lengths of the two legs forming the right angle of a right isosceles triangle, we can utilize this ratio to determine the length of the third side. This ratio enables us to make a more rapid calculation of the length of the hypotenuse once we have determined that we are dealing with a right isosceles triangle and to divine the answer as though we were telepathically gifted. So if each leg of a right isosceles triangle is equal to 5 feet in length, then we know that the hypotenuse is equal to $5\sqrt{2}$, which is equal to 5 multiplied by the square root of 2 (e.g., 1.414213562), or 7.071067812. Most people will not, of course, bother to carry out their calculation to such a level of precision and will usually be happy to write 7.07.

By this point, you may have hit upon one of the most important features of trigonometry—the angles determine the lengths of the legs and hypotenuse and the lengths of the legs and hypotenuse determine the angles. Once you appreciate this concept, you will be well on your way to mastering the basic principles of trigonometry. This concept will also help to provide some guidance when you are trying to solve a particularly thorny trigonometric problem because it underscores the interrelationship among the various elements of the triangle; knowledge of one element (length) will facilitate knowledge of another element (angle).

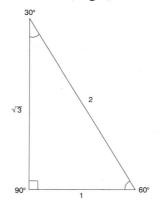

FIGURE 2

Before continuing our discussion of trigonometry, we should consider another very common right triangle shown in figure 2—the so-called 30°/60°/90° right triangle, which, despite its cumbersome name, is perhaps even easier to grasp than the right isosceles triangle. This triangle consists of three sides, the shortest of which is half the length of the hypotenuse. The three sides of this triangle can also be expressed in terms of the ratio 1:$\sqrt{3}$:2. As with the ratio for the right isosceles triangle, this ratio boasts another square root sign. If we calmly and deliberately go back to the Pythagorean theorem, we see that this is a very simple problem. After all, $1^2 + ?^2 = 2^2 = 4$? As 1 multiplied by itself is still equal to 1, we know that the other side, when squared, must be equal to 3—so the only number that can be squared (multiplied by itself) to equal 3 is $\sqrt{3}$. Given these two lengths, the hypotenuse must be equal to 4.

Certainly we can all appreciate the wondrous symmetry of these ratios and the magical ease with which the Pythagorean theorem can enable us to seize any right triangle and squeeze out secrets about the lengths of its sides, but we might want to know how all of this information can be put to practical use. In short, how can we use our knowledge of trigonometry for practical ends?

We can use the ratio offered by the 30°/60°/90° triangle to solve problems relating to the heights of objects. How so? you may ask. Suppose I am designing an office building in the downtown business district and several of my prospective neighbors are complaining about the shadow that my building will cast. The length of the shadow cast by my building will be dependent on its height and the angle at which the sun strikes it. We might further assume that the really vocal opponents of my project are located about 500 feet away from my building site. Accordingly, I might want to calculate the height of the building I could construct that would cast a shadow of no more than 500 feet when the sun is at a 30 degree angle in the sky (its angle of elevation). I would use the ratio of the 30°/60°/90° triangle (1:$\sqrt{3}$:2) to set up the following expression: The height to which the building should be built is to the length of the shadow as

1 is to $\sqrt{3}$ or $x/500 = 1\sqrt{3}$. If we multiply both sides by 500, we get a value for x equal to $500\sqrt{3}$, which is roughly equal to about 288.68 feet. So a building less than 289 feet in height will cast a shadow 500 feet in length when the sun is 30 degrees above the horizon (presuming we live in a comparatively flat area of the country).

This particular equation can be used to calculate the heights of all kind objects. Suppose you are visiting a big city and you decide to calculate the height of a marble column in the park that was erected to honor the city's mayor. You look up in the sky and estimate that the sun is about 30 degrees above the horizon. Then you pace out the length of the shadow cast by the column and find that it is about 34 feet in length. Then you recall the 30°/60°/90° ratio $1{:}\sqrt{3}{:}2$ and remind yourself that the height of the column is to the length of the shadow as the ratio $1{:}\sqrt{3}$. So we go through the same process as before and set up the equation $x/34 = 1\sqrt{3}$. By multiplying both sides of the equation by 34, we get a value for x equal to $1\sqrt{3} \times 34$, which is roughly equal to 19.3 feet. So you can therefore conclude that the column is really not much of a memorial at all—being less than two dozen feet in height—and barely extends up above the tops of the passing city buses. But now you have acquired some knowledge through your study of trigonometry that will enable you to look at and solve certain types of problems involving calculations of spatial dimensions.

THE WORLD OF TRIGONOMETRIC FUNCTIONS

Now that you have learned firsthand how a little knowledge about the ratios of triangles can assist you in carrying out certain basic calculations, we need to learn an "f word"—function. We will encounter functions throughout the world of mathematics, but trigonometry offers a unique set of functions that can enable us to determine the lengths of the sides of triangles or the distances separating objects that are placed on opposite endpoints (vertices) of a triangle. If we know the angles of the triangle and just the length of one leg of the triangle, for example, then we can determine the lengths of the other two legs of the triangle. This skill can be important to people who must calculate distances in their daily work—whether they are astronomers or surveyors. But before we get to the ways in which astronomers have determined the distance to the moon or the sun, we need to talk a little bit about these trigonometric functions.

Trigonometry has six functions that are known as the sine, cosine, tangent, cosecant, secant and cotangent. Too few people have any idea as to what these terms actually mean. But these seemingly imposing terms are nothing more than the six ratios that can be formed by the six possible combinations of any two of the three sides of a right triangle. That sounds fairly simple, to be sure! If A, B, and C designate the three sides, then we can have six ratios: (1) side A to side B (sine), (2) side A to side C (cosine), (3) side B to side C (tangent), (4) side B to side A (cosecant), (5) side C to side A (secant), and (6) side C to side B (cotangent). The ratios are not going to be fixed values because they will vary with the measures of the angles of the triangle. But the good news is that we can use our knowledge of these ratios to help us solve many different types of problems in trigonometry.

For purposes of illustration, we will use a few examples to illustrate the ways in which three of these ratios—the sine, the cosine, and the tangent—can be used to help determine the length of a particular side of a triangle. This is not some sort of theoretical exercise because the use of trigonometric functions has many immediate real-world applications, particularly in the construction of massive engineering projects.

Most of us have driven through tunnels at one time or another and not given a second thought to the degree of technical expertise needed to complete such a project. But the technical skills needed to construct a first-rate tunnel—particularly those tunnels extending for great distances through mountains or under rivers—necessarily require a basic knowledge of trigonometric functions. The Greek mathematician Heron pointed out that trigonometry could be used to enable two crews working from opposite sides of a mountain to meet in the middle. Now bear in mind that this is an extremely helpful tool because people digging tunnels, particularly in the ancient days when the lives of the workers themselves were not always considered to be of great importance, had no way to see where they were going. The absence of telephones and other wireless communication devices prevented them from talking to each other to ensure that their respective tunnels would actually meet in the middle. But Heron realized that he could solve this problem if he could draw a sufficiently large triangle in which the two crews perched on opposite sides of the mountain could be placed at points A and B (see figure 3). The third point C on the triangle in figure 3 needed to be sufficiently far away so that the angle ACB would be a right angle and an observer standing at C could see both crews simultaneously—even though they would not be able to see each other.[3]

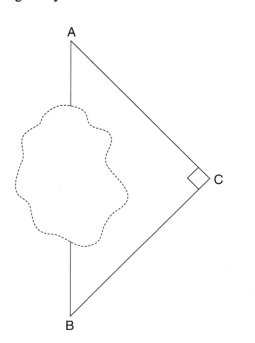

FIGURE 3

The observer could then play a valuable supervisory role and make sure that no one was slacking off too much outside either tunnel. But Heron had other ideas beyond those relating to extracting the maximum amount of work from tunnel-digging slaves. He realized that he could use trigonometry to calculate the paths that the two digging crews would follow so that they would meet each other in the middle of the mountain. By setting the observer's angle C equal to a right angle, the observer would measure the distance from C to B and the distance from C to A. Once he knew the lengths of the two sides of this imaginary triangle, he could then calculate the tangent of both angles (tangent A = BC/AC and tangent B = AC/BA), which would enable him to calculate the directions that the crews should take to meet each other in the middle. Moreover, the information provided by these calculations would enable him to determine the length of the path between B and C. This calculation would be particularly helpful in preventing a mining crew with inept crew bosses from urging the diggers onward in meandering paths throughout the interior of the mountain as they would know the approximate distance at which they should be running into the other crew.

One can only imagine the enormous savings in labor costs that trigonometry brought to the mining industry. But this savings underscored the enormous practical value of trigonometry—which has been replicated in numerous industries throughout modern history. Perhaps the most profound impact of trigonometry was more philosophical than practical: It was used by both Hipparchus and Ptolemy to measure the sizes and distances of some of the celestial bodies such as the sun and the moon, thereby greatly expanding the horizon of the known universe.

You would think that Hipparchus would have been faced with an almost insurmountable task in trying to calculate the distance from the earth to the moon. After all, the early Greeks did not have any means to physically measure the distance to the moon. But Hipparchus was not one to give up easily, so he returned to the basic trigonometric principles that had been used earlier by Eratosthenes to measure the circumference of the earth. Eratosthenes had found that when the sun was directly over the Egyptian city of Syene at high noon so that he could see down to the bottom of a well, a person standing in the city of Alexandria about 500 miles to the north would find that the sun at that very same time cast a shadow of about 7½ degrees. Eratosthenes realized that if he divided 360 degrees by 7½ degrees, then he would get the proportion of the circumference of the earth represented by the distance from Alexandria to Syene—about ⅟₄₈ of the total circumference. Once he determined that it would take about 500 miles of travel to cover ⅟₄₈ of the circumference of the earth, then it was easy to multiply 48 by 500 to determine that the earth was about 24,000 miles in circumference. This figure is remarkably accurate and not too far off from our modern estimates of 24,900 miles around the equator. Indeed, the only persons who were never overly impressed with the accuracy of Eratosthenes' calculations were some of the individuals who still believe that you can fall off the edge of the earth if you sail beyond the horizon.

Hipparchus believed that Eratosthenes was definitely on the right track in using geometric principles to determine the circumference of the earth. But in trying to calculate the distance to the moon, Hipparchus needed to modify the approach used by Eratosthenes. It is doubtful that Hipparchus's work was motivated by a desire to travel to the moon. Indeed, it seems that his interests were primarily rooted in a desire to understand better the physical universe around him, which at that time was thought to be little more than the earth surrounded by a canopy of stars, planets, and moons. Few people gave very much thought to whether there was anything beyond this "firmament" as they were embroiled in the ongoing wars between the various Greek city-states. But for those few men who were given to contemplation and who could appreciate the subtleties of nature, the universe proved to be an irresistible subject and offered a variety of tantalizing puzzles for their consideration. For Hipparchus, the challenge was to gain some understanding of the universe by trying to ascertain its magnitude. And an important step in this process was to figure out the distance to the moon.

So Hipparchus selected the point C at the center of the earth from which he could extend an imaginary line through the surface of the earth at point S up to the moon at point M (see figure 4).

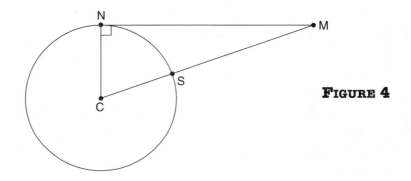

FIGURE 4

This line was bounded at one end by C and at the other end by M. Hipparchus then selected a second point on the surface of the earth N that happened to be that point at which the moon would barely be visible over the horizon when it was directly overhead at S. Are you still with us? Hipparchus then drew a second line from point N up to the moon at point M. Finally, he connected C and N together with a third line, thus completing his triangle shown in figure 4.

Hipparchus had drawn a triangle, which was not something that in and of itself would challenge the artistic skills of even most grade school children. But Hipparchus knew from the work of Eratosthenes that the distance between C and N (the radius of the earth) was about 4,000 miles because the diameter of the earth itself had been determined to be about 8,000 miles. He also knew that if he used a right triangle with the 90 degree angle at point N, the remaining two angles at C and M would also be equal to 90 degrees. The challenge was to determine the measure of angle C, which Hipparchus calculated by measuring the arc that one would have to travel in going from N to S. Although it was by definition less than 90 degrees because N, not C, was the right angle in this triangle, it was not much less—measuring 89 degrees, 4 minutes, and 12 seconds.[4] (Remember that 1 degree consists of 60 minutes and 1 minute in turn consists of 60 seconds.) By simple subtraction, this meant that the remaining angle M was equal to a mere 55 minutes and 48 seconds—a little less than 1 degree.

Having determined the angle C by his careful measurement of the arc from N to S, Hipparchus then brought trigonometry into the fray. He calculated the cosine for angle C by dividing the segment NC (which, you will recall, had been determined to be 4,000) by the segment CM. The cosine for C was thus determined to be equal to 4,000/CM. Hipparchus then determined that the cosine for the value 89 degrees, 4 minutes, and 12 seconds was 0.0163 (a trigonometric ratio calculated by Hipparchus) that was in turn equal to 4,000/CM. He could then multiply both 0.0163 and 4,000/CM by CM to get 0.0163CM = 4,000. By dividing both sides of this equation by 0.0163, he obtained a value for CM of 245,400. As a result, Hipparchus concluded that the moon was nearly one-quarter of a million (about 245,000) miles away from the planet earth. Needless to say, this result stunned many people who had thought that the moon was much closer—perhaps just a little bit beyond the next mountain range.

But Hipparchus did not stop with the moon as he was able to use the same method to calculate the distance from the earth to the sun. The triangle used in figure 4 would be the same but the point M would be replaced by a different point P to represent the sun (see figure 5). The calculation of the angle of C in this revised model would be a little different because the angle here would be even closer to approximating a right triangle, measuring about 89 degrees, 59 minutes, and 51 seconds. This leaves the third angle P to be equal to about 9 seconds of arc, that is, a little less than ⅙ of a minute of arc or about ⅟₄₇₀ of a degree of arc. The measure of angle was far smaller than that arrived at in the calculation of the angle of arc of point M when Hipparchus was trying to calculate the distance to the moon—by a factor of about 370.

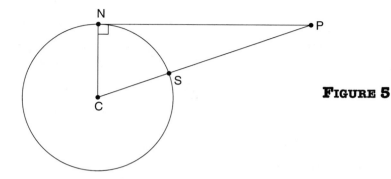

FIGURE 5

The angle for C is larger and the corresponding angle for P is much smaller because the sun is much further away from the earth. The angle P in figure 5 is about ⅟₃₇₀ the size of the angle M in figure 4. Not coincidentally, the distance of the sun from the earth is about 370 times as great as that of the moon from the earth. If we multiply the distance of the earth from the sun (245,400 miles) by 370 (the approximate magnitude by which the angle M in figure 3 exceeds the angle P in figure 4), then we obtain a figure for the distance to the sun equal to almost 91,000,000 miles. This is a little less than the commonly accepted figure of 93,000,000 miles because we rounded off some of our numbers in making the comparisons between the two distance models. But this calculation underscores the usefulness of trigonometry in calculating distances between objects in outer space as well as the sheer mathematical brilliance of some of the early Greek intellectuals. Mathematicians also used these very same principles to determine the diameters of celestial bodies such as the moon and the sun. Needless to say, these measures obliterated the ancient visions of a tiny, self-contained canopied universe and hinted at the vastness of the cosmos that would later become unveiled with the introduction of telescopes and other modern astronomical instruments.

We all owe a great debt to Hipparchus who labored for much of his life making the tedious calculations of the trigonometric ratios (such as the cosine for the angle C in figure 3) that are contained in tables found at the backs of most mathematics textbooks. Indeed, the accuracy of these calculations was so great that modern mathematicians have not been able to greatly improve upon them. These tables have also made it possible for mathematics students to concentrate on solving the equations themselves instead of having to spend so much of their available time calculating these trigonometric ratios. Hipparchus certainly made it possible for many of us to enjoy more of our leisure time than might have otherwise been possible if we had been required to calculate these very same trigonometric ratios by ourselves. In doing so, he laid the groundwork for the modern study of trigonometry and the many discoveries and technological innovations that would be made by applying its basic principles.

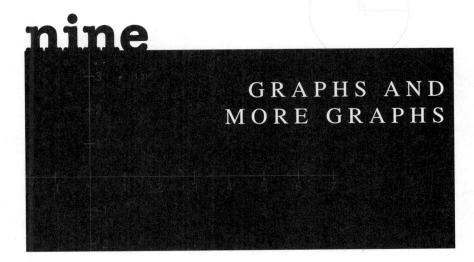

nine

GRAPHS AND MORE GRAPHS

Some mathematicians like to think of themselves as artists who use numbers and equations in the same way that painters use oils or watercolors. We sometimes think of mathematics as consisting exclusively of elaborate equations. Graphs also play an important role in mathematics because they help to give us "snapshots" of quantitative relationships that may exist between two or more variables.

We may be surprised at the extent to which mathematicians resort to graphs to illustrate changes in different values or quantities. Why would they not simply prefer to rely on their charts of numbers? Quite simply, the graph provides a very explicit description of the relationship between two or more variables that can usually be comprehended very quickly. When dealing with many dozens or hundreds of measurements, the mathematician can become bogged down in the numbers and lose sight of the ultimate relationships that these numbers may describe. To avoid this morass, each of these measurements may be plotted on a graph and the mathematical relationships suggested by these measurements taken in with a single glance.

Although this discussion of graphs, variables, and relationships may sound very academic, it is actually quite simple and can be easily understood. When we talk about trying to understand the relationship between two variables, it may seem as though we are discussing something unworldly and esoteric when in fact we are talking about something we all do each day. We all encounter situations in which we see one variable having an affect (positive or negative) upon another variable. In other words, a change in one variable causes some sort of change in another variable.

Suppose that you are in an automobile driving along the freeway, for example, enjoying the wind blowing through your hair and periodically testing the acceleration of your engine. Whenever you need an extra burst of speed, you press down on the accelerator. Your pressing down on the accelerator causes your automobile's speed to increase. The more you press the accelerator, the faster your automobile races down the highway. As one who is particularly alert to the subtle workings of the physical world, you also know that the speed of your automobile will decrease if you ease up your pressure on the accelerator. So you will find yourself hurtling faster and faster as you depress the accelerator further toward the floor whereas you can expect the opposite result if you remove your foot from the accelerator.

The fact that the speed of the car changes as you depress or release the accelerator provides a very nice example of the way in which changes in one variable (the extent to which the accelerator is depressed) affects changes in another variable (the speed of the vehicle).[1] Indeed, we could construct a graph to illustrate this relationship between the amount of acceleration and the speed of the automobile (see figure 1). We could have the horizontal axis (the *x*-axis) reflect the speed of the vehicle in miles per hour and we could mark the vertical axis (the *y*-axis) with tenths of inches (0.5 inches, 1.0 inches, and so on) to show the extent to which the accelerator is being pushed downward.

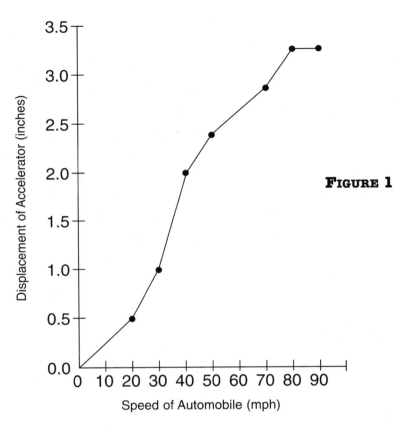

FIGURE 1

This is not a very mysterious process because we merely have to connect the points that we plot on this two-dimensional graph—which is called a Cartesian plane. This graph, which is named after the famed French philosopher and mathematician, René Descartes (whose work is discussed in more detail in chapter 13), consists of two perpendicular lines that intersect at a point of origin that is customarily assigned the value of zero. The horizontal line is the *x*-axis and is marked with positive integers (1, 2, 3, . . .) to the right of 0 and negative integers (−1, −2, −3, . . .) to the left of 0. The vertical line that intersects the *x*-axis at the point of origin is the *y*-axis and is marked with positive integers (1, 2, 3, . . .) above the 0 and negative integers (−1, −2, −3, . . .) below the 0.

We could then get on the highway, ruler in hand, and begin measuring the distance we pressed the accelerator downward at any given speed. No doubt we would soon discover that the more we pressed the accelerator, the faster we moved along the road. We would begin plotting points on our graph that would relate the amount the accelerator was depressed to the speed of the vehicle at a given time. Our chart would begin to take shape and we would, in all likelihood, gradually construct a line that supports our initial assumption about a heavier foot on the gas leading to greater speed. This is not the type of experiment that would win a Nobel Prize because, like the papier-mâché volcano at many elementary school science fairs, it has simply been done too many times. But it does enable us to conclude, after plotting a sufficient number of points on our graph, that there is a direct relationship between the independent variable (the extent the accelerator is pressed) and the dependent variable (the speed of the vehicle). Our graph will reveal that the harder we press down on the accelerator, the faster our vehicle will travel. This graphical representation will appear in the form of an upward sloping line that, despite this apparent direct relationship, will not continue forever. It will instead reach a plateau because we can only press an accelerator down only so far before it hits the floor and will go no further. Similarly, your car, no matter how high the octane of the gas you put into it, can only go so fast before it reaches a maximum speed. This means that any direct relationship represented by an upward sloping line will eventually flatten out because the speed of the vehicle cannot continue to increase without limit.

Another example of the relationship between two variables may be demonstrated by any one who has enjoys gardening. Suppose that you believe that if a little fertilizer on roses is a good thing, a lot of fertilizer on roses is a wonderful thing. You could verify your hypothesis by selecting a few of your hardier bushes and dousing them with differing amounts of fertilizer. You could construct a graph with a horizontal axis showing the amounts of fertilizer applied to the individual bushes and a vertical axis showing the increases in the height of each bush. You could have a separate graph for each plant or you could use different colors to plot out the points for each plant on a single page of paper.

So we might track the growth progress of six different rose-bushes, all of which appear to be about the same height at the beginning of our experiment. We would want to make sure that they were all the same kind of rosebush because it would not be scientifically useful to use a number of different varieties of rosebushes that might have very different rates of growth. The validity of our experiment would be dependent upon our "comparing apples against apples." It would not add very much to the body of knowledge of roses if we were to plant three rapid-growth rosebushes and three slow-growth rosebushes and then give them all equal doses of fertilizer. After all, the rapid-growth rosebush might grow more in a month than the slow-growth bush would grow during an entire year. The differing growth rates among our bushes could then hardly be said to be dependent upon the fertilizer being applied. After all, you could plant the slow-growth rosebushes in a tub of pure fertilizer and it might still have no meaningful possibility of matching the growth rate of the rapid-growth rosebush. In fact, the slow-growth rose might even die because it might not be able to tolerate the extremely toxic environment of a pure fertilizer soil. So this experiment would not enable us to demonstrate our original point relating to the interaction of the independent and dependent variables. In this case, the independent variable would be the amount of fertilizer we dropped on a rosebush and the dependent variable would be the growth of the bush itself. There would presumably be a direct relationship between the amount of fertilizer dropped on the bush and the height it reached in a given amount of time. As a result, our graphs would probably rise upward moving from left to right to reflect the greater height in the bushes (vertical axis) that was obtained by increasing the amounts of fertilizer (horizontal axis). We would also expect that the lines on the graphs for the bushes that received comparatively heavy amounts of fertilizer would rise more sharply, thereby reflecting the increased rates of growth. But we could also surmise that this relationship would not continue indefinitely. If we continued to increase the dosages of fertilizer we would not necessarily continue to see progressively greater rates of growth. It is more likely that we would reach a point of diminishing returns in which the increased toxicity of the heavy doses of fertilizer would begin to retard the rate of growth of our bushes. So the upward curve of our graph would level off at some point because our bushes would reach a stage at which they would simply be unable to absorb any additional fertilizer. This is not the end of the story as we pointed out above because if we continued to add more fertilizer, we would see all sorts of unpleasant things take place, including the deaths of the bushes themselves. But this experiment would still have value in a mathematical sense because it would underscore the relationship between the independent and dependent variable. The important point to remember is that the assignment of values to the independent variable necessarily determines the value of the dependent variable. This point is of immense importance in algebra because it is this technique that makes it possible to solve the unknown values in a given equation.

Algebra books have thousands upon thousands of equations that are in turn replete with alphabetic variables such as x, y, and z. This is not a quirk of fate because algebra in its most fundamental form involves the manipulation of symbols and numbers to obtain solutions to equations. But the fact that letters are used to stand for unknown numbers somehow seems to intimidate many students who are new to the algebra experience. There is something about letters and numbers appearing side by side in the same mathematical equation that is slightly upsetting to some of us who are used to keeping our numbers and letters apart. However, once we accept the fact that the letters appearing in algebraic expressions are nothing more than numbers wearing masks, we should be able to breathe easier. It may help to bear in mind that numbers and letters are essentially arbitrary symbols that were assigned specific meanings by their inventors; there is nothing inherently mystical or magical about numbers and letters themselves.

When you start fiddling with algebraic equations, you must eventually try to solve those equations. As mentioned earlier, one of the most common situations that we run across is equations containing at least two different variables such as x and y. In other words, we might be given an equation such as $2x + 1 = y$ and asked to solve the equation. We need to recall our earlier observation that algebra is essentially a discipline in which symbols are manipulated in order to obtain solutions. But the degree of manipulation required for solving an equation such as $2x + 1 = y$ is very limited. Indeed, our analysis of the equation involves little more than determining a value for x and then solving the equation for the value of y. This means that we substitute a given number such as 1 for x. This gives us the following equation: $2(1) + 1 = y$ or $2 + 1 = y$.

Now we are faced with the truly daunting task of trying to think what number can be substituted in place of y to solve this equation. Despite the intricacies of the new math, we would eventually determine that $2 + 1 = 3$ because the only value for y that will solve the equation is 3. As a result, we would obtain (1, 3) as an ordered pair where the first number stands for a value of x and the second number stands for a value of y as a solution set for the equation $2x + 1 = y$. But this is by no means the only solution that we can trot out for this particular problem because we can substitute any number for one of the variables (e.g., 2, 5, 26, 91, and so on) and thereby create any number of solution sets. What is the point to such an approach? Well, it makes it possible for us to create a graphical description of this equation by plotting the points described by these various ordered pairs. If we want to locate the point on this plane described by the pair (1, 3), for example, we would start on the x-axis at the point of origin and move 1 unit to the right. We would then move upward 3 units along the y-axis and then place the point that is located 1 unit to the right of the unit of origin of the x-axis and 3 units above the unit of origin of the y-axis. We would also want to make sure that we have our negative and positive values for our positions correct because whether our x-value or our y-value or even both values have negative values will result in their being positioned in different quadrants of our Cartesian plane (see figure 2). At the risk of seeming somewhat redundant, we can look at the particular quadrant in which the point is located and conclude whether either one or both of the values for x and y are negative.

We have been dealing with the upper right quadrant of our Cartesian plane up to this point where all the numbers are positive integers. But things start to change if we move to the left side of the point of origin of the x-axis. Although our y-values are still positive, our x-values are now negative numbers (see figure 2). And if we move below the x-axis, then both our x- and y-values will become negative.

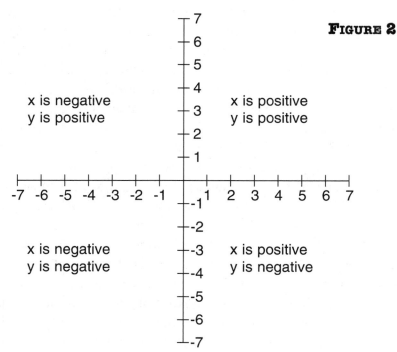

FIGURE 2

This revelation may or may not be of concern to you, depending in large part on your attitude towards negative numbers. Some people have a conceptual difficulty dealing with negative numbers because they believe that such numbers must have some "other-worldly" qualities not possessed by the ordinary, run-of-the-mill positive numbers. We can relate to the numbers 8 or 10 because they can be readily associated with tangible objects such as a collection of 8 apples or 10 tires. But we feel stymied when we deal with negative numbers because we cannot associate them with "negative" objects. We would have no way to conceptualize negative apples or negative tires because these objects do not exist. But this is the wrong way to look at negative numbers because we need to bear in mind that negative numbers are arbitrary symbols that were devised by mathematicians as a natural extension of the counting exercise and, hence, the number line itself. There is nothing magical about them because they were not developed to represent "negative" objects but merely to provide continuity in the counting process. But the idea of negative numbers is not so obscure as we might first think because all of us who have ever bounced a check know full well the meaning of a negative bank account balance. We have a negative account because we owe the bank the amount of our shortfall (assuming that the bank bothered to pay the check in spite of the overdraft) and not because we have "negative" money. This idea of a bank overdraft is thus analogous to wandering through the negative part of a number line or the negative side of the x-axis.

To return to our discussion of the Cartesian plane, we see that a point located on the upper-left quadrant in figure 2 has a negative x-value and a positive y-value. But if we move to the lower-right quadrant, we find that we have a positive x-value and a negative y-value. Finally, a shift to the lower-left quadrant leaves us with negative values for both our x and y coordinates. We have thus tried to familiarize ourselves with the basic characteristics of a Cartesian plane so that we will have a clue or two as to the ways in which positive and negative values relate to each other in this two-dimensional coordinate system.

How did Descartes stumble upon this coordinate system? As with most important contributions to the sciences, it was reportedly inspired by seemingly mundane observations. In Descartes's case, he was laying in bed watching the branch of a tree shift back and forth across the glass panes of a window. This observation prompted Descartes to think of the windowpane as a field with crosshatched horizontal and vertical coordinates. But Descartes realized that he could specify the location of the branch anywhere on the windowpane by selecting a corner of the window as his point of origin and then counting the number of squares over and the number of squares above the point of origin to reach the desired position. This is not to say that Descartes spent all of his time in bed watching tree branches scrape across the window or that anybody else who happened to be passing through his bedroom was content to engage in the exercise of calculating the displacement of the tree branch. But it does illustrate the point that a seemingly simple event in nature can provoke thoughts that can ultimately have very profound consequences in the development of our applied knowledge.

Now that we have thoroughly immersed ourselves in the historical roots of Cartesian geometry, we can return to the idea of actually graphing algebraic functions on a Cartesian plane. One of the advantages of modern mathematics is that we do not need a windowpane and a tree branch to carry out such an operation but can get by with a pencil and paper or, alternatively, a graphing calculator. The advantage of the pencil and paper, of course, is that it does not require a battery in working order. The use of pencil and paper is a slower, more cumbersome process. And since this process is best explained in the traditional pencil and paper form, we shall proceed accordingly. Those persons who are fortunate enough to have a graphing calculator at their disposal can proceed ahead at their own risk.

As far as graphing our algebraic functions is concerned, we

shall use the equation $2x + 1 = y$ for which we have already obtained the solution set $(1, 3)$. We had initially plugged the value $x = 1$ into the equation that, when solved, yielded a y-value of 3. Of course we are not slaves to the mathematical fashion world and could carry out this solution process in the reverse order by picking a value for y and then solving for x. Some people who read from left to right find it more comforting to start with the y-value and then solve for x and there is nothing wrong with using such an approach in a democratic society. Because x precedes y in our alphabet we can go with the flow and pick some more values for x and solve for y in order to create a series of solutions sets that will facilitate our creating a graphical representation of the solutions for this equation.

Now you might say that this should not be very much of a problem because you recall from your basic geometry that two points define a straight line. As a result, you may believe that this exercise in graphing functions will be a piece of cake because you only have to come up with another x-value and solve for y and then you can be well on your way toward completing the task at hand. But there is no guarantee that two solution sets for our equation will necessarily give us a completely accurate picture of this function. In the case of this particular equation, $2x + 1 = y$, we first plot the point $(1, 3)$ on our graph (see figure 3).

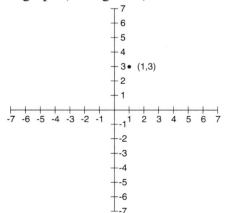

FIGURE 3

Now a graph with only a single point is not very useful in terms of illustrating the various solution sets for the function even though it is very neat and clean. Graphs having a single point are certainly uncontaminated by the lines that run helter-skelter over some of the more complex graphed functions, but they also have their limitations. As a result, we need to select several other values for x in order to determine the true "shape" of our function on our Cartesian plane. We would probably prefer to pick a few whole numbers that can be used to solve our equation with a minimal amount of effort. This means that we might choose to shy away from fractions, square roots, and other mathematical qualities that may not be as easily employed in the solution of our equation. We will also assume that no more than five separate solution sets will be required for the plotting of a suffi- ciently accurate graphical function. As we have already used $x = 1$ for our first solution set, we cannot use it again because we would merely be placing the same point on top of the first point over and over again. While we would have a very distinct and obvious point marking this solution set, we would have little else to show for our effort. We would probably prefer to use different values for x so that we might actually get somewhere in drawing our graph. As a result, we might select 0, 1, –1, 2, and –2 as our group of x-values to be applied to our equations and graphed on our Cartesian plane. This merely requires that we repeat the process we used before and plug these numbers into the equation. As shown below, this operation gives us the following values for x and y.

TABLE 1

x	y
0	1
1	3
–1	–1
2	5
–2	–3

Once we plot each of the five points on our graph, we obtain the function described by these five coordinates (see figure 4).

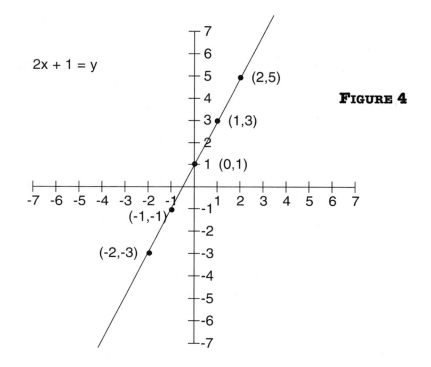

2x + 1 = y

(2,5)

(1,3)

(0,1)

(-1,-1)

(-2,-3)

FIGURE 4

We are clearly making progress because we have nailed down the concept of the coordinate system and the idea of transferring numerical representations to a graphic description on a two-dimensional graph. But this exercise also illustrates that many algebraic equations can have a virtually unlimited number of solutions as underscored by the continuous line traced out on figure 4. We may now wander into the neighboring domain of geometry where all sorts of things can be learned about points and lines and planes and spaces. In particular, we need to consider the idea of a point, which geometers tell us is the most elementary concept of geometry. In a sense, the point is very accessible to us because we can grab a pencil with a very sharp end and press it against a piece of paper to create a point. Regardless of whether we are dealing with one point or many, however, we must still try to grapple with the fact that the point, though easy to visualize, is actually very profound in a mathematical sense.

A point, according to mathematicians, is dimensionless because it has no length, width, or depth. The usual way to try to understand this idea is to imagine a writing instrument such as a fountain pen with an infinitely fine nub, much finer than any other fountain pen ever created. Because the nub is so precisely engineered, it can make a far smaller mark when put to paper. Now in the real world we know that any mark we make, no matter how carefully placed, will have dimensions of some sort because the point itself has a certain length and width, even though it may be a minute fraction of a millimeter. In the theoretical world, however, a fountain pen with an infinitely precise nub could make an infinitesimal point lacking any spatial dimensions at all. Although there is not much demand for an infinitely fine nub in the real world, it does help to illustrate the classic definition of a geometrical point.

Because the point has no dimensions, an unlimited number of points can be squeezed into an automobile or a teapot or a thimble. Indeed, we could pile an infinite number of points onto a grain of sand or a speck of dust. What many people find troubling when dealing with the geometrical point is the idea that an infinite number of points can occupy the biggest star or the smallest atom and that both sets of points are equivalent to each other. If this is not enough to cause amazement, then try to imagine that there are an infinity of points between any two of the points in either of these two sets. We may thus assert with great confidence that there are an infinite number of points between any two points in the biggest star and an infinite number of points between any two points in the smallest atom.

To return to our original statement regarding the continuous line that encompasses the solutions for our problem $2x + 1 = y$, this understanding of the dimensionless nature of geometric points is crucial because we must now grapple with the concept of the line. A line is defined by our geometer friends as extending infinitely far in both directions. Not surprisingly, a line contains an infinite number of points. But every line contains the same number of points because every line by definition is unending. Lines give us far more flexibility in our graphical presentations because they contain an infinity of points and move off in both directions without limit. A continuous line, such as the one graphed in figure 4, also contain an infinity of points and each point represents a solution to the equation $2x + 1 = y$. As we move along this line, we move from one point to another, each of which offers both an x-value and a y-value that, when plugged into our equation, can be used to obtain a solution. The fact that this function is a continuous line shows that there are an unlimited number of ordered pairs that satisfy this equation. In dealing with these ordered pairs that can be used to satisfy this equation, we also need to familiarize ourselves with the proper language that mathematicians like to use in solving such equations. When we select a value for x and then determine the value for y, we say that y is a function of x. This technical language basically means that we are arbitrarily selecting values for x (our independent variable) that in turn will determine the values for y (our dependent variable). The independent variable can be any value we choose although we do not have complete liberty to substitute anything we want. However, if we substitute a numerical quantity for x, we will instantly determine the value for y. Hence, y is not as free spirited as x because y's identity is inexorably determined by x's identity. In short, dependent variables have no autonomy. It is not a job for rabble-rousers, revolutionaries, or those who cherish the concept of free will.

Now you must no doubt be wondering whether your artistic needs can be satisfied by plotting such simple algebraic solutions. After all, the graph that is plotted by finding ordered pairs for the equation $2x + 1 = y$ is a straight line, which is not really the most demanding picture for an artist to paint. As we wander through the landscape of algebra, however, we encounter all types of functions, some of which will yield very intricate functions that, when plotted upon a diagram, will have all sorts of curves that will tempt even the most creative artists. How much graphing fun might we be able to expect from these more complex functions? Well, we could start with a very simple example such as $x^2 = y$ and plug in several values for x to get the ordered pairs that will yield a very nifty graph. So let us pick our standard x values (–2, –1, 0, 1, 2), which gives us the following ordered pairs: (–2, 4), (–1, 1), (0, 0), (1, 1), and (2, 4). If we look at figure 5, we will see that we are no longer dealing with a straight line but instead a very lovely symmetrical function that curves downward from the upper-left quadrant, passes through the point of origin where both the x- and y-axes intersect, and then swings back upward into the upper-right quadrant. This function continues upward indefinitely in both the upper-left and upper-right quadrants, meaning that there are any number of ordered pairs that will serve as solutions to this problem.

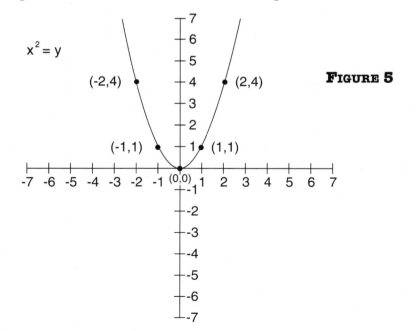

FIGURE 5

Now it is time to add to our mathematics vocabulary by learning two new terms that relate to functions: *domain* and *range*. The domain is the set of permissible values of the independent (x) variable in a given function, which in this case shall be defined as $y = x + 1$ (where x is equal to 2, 5, 8, 9). So if we have a function for which the numbers 2, 5, 8, 9 are the only values that may be substituted for the independent (x) variable in an equation, then we would call these four terms the domain of that equation. The range of a function is the set of values of the dependent (y) variable that would correspond to the values substituted for the independent (x) variable. So if substituting the numbers 2, 5, 8, 9 for x in our mystery equation gives us the corresponding values of 3, 6, 9, 10 for our dependent (y) variable, then we have described in one fell swoop a very compact set of independent and dependent variables.

The graphed function for our equation is fairly simple because it is an upward- or positively sloped line that has certain boundaries along both the horizontal and vertical axes of our graph due to the restriction we placed on the permissible values of x. This is all very exciting news because it enables us to avoid the unnecessary detail of graphing those values for x and y that are beyond the boundaries of our domain and range. But we should also point out that the continuous line connecting the four sets of ordered values has nothing to do with the function itself but is merely a very descriptive way to show the pattern created by linking together the positions described by our equation. As far as the mathematical significance of the line is concerned, however, we know that it is of no relevance to our particular function because we have limited ourselves to but four specified values for x. The line would be relevant only if $y = x + 1$ for all values of x, including nonwhole numbers. So we would have more than four points along this line because we would also be plotting all sorts of fractions, decimals, and other numbers that would lie between our four sets of ordered pairs on our graphed function. If we graphed all possible values of x between the boundaries of this function (where x is equal to or greater than 2 and equal to or less than 9), then the line would reflect the fact that this function is a continuous function with an infinity of different values for x between these boundaries.

But you are no doubt concerned about all this loose talk relating to dependent and independent variables. After all, how does one go about selecting an independent variable? Would you be able to decide for yourself whether a variable should be tagged as being independent or dependent? The question is of an even more fundamental nature in that we must consider situations where we must decide which variable depends on the other—even though it is not always so clear as to which direction this dependency may run. Suppose, for example, your mathematics teacher gives you a problem in which he asks you to decide which is the dependent variable: (1) the time it takes you to reach your school or (2) the speed at which your bus travels through the city to your school. You must admit that this question has something of the chicken and egg circularity to it. After all, you are not told which variable is dependent upon the other and this lack of information might be confusing. Here, we need to determine the issue of dependency by putting either statement into a dependent and an independent context. Even though this point may have a certain arbitrariness to it, the distinction must be made so that mathematicians can better appreciate the causal relationships between independent and dependent variables.

If we take our two variable statements relating to the travel time of the bus and the speed of the drive itself, then we can come up with two very different statements: "The time the bus trip requires depends on the speed it travels" or "The speed at which the bus travels depends on the time the trip takes." Which sentence makes more sense from a logical point of view? We must rely on our own common sense and look at the sentences themselves. The first sounds like a fairly reasonable statement, whereas the second leaves us with an uncomfortable feeling that it is too undefined or too "mushy." The first sentence seems more distinct, more defined, and more understandable. The second sentence is more unsettling because the idea that the time the trip takes somehow determines the speed at which the bus travels is very strange and difficult to understand. The first sentence, by contrast, is easy to grasp in that we are not distressed at having to consider the time a trip takes as being a function of the speed at which the traveler travels. Indeed, this is a very common way to look at any trip on the highway because we base our travel speeds on several things, including the estimated time the trip will take at a given speed. If we were to accept the first sentence, we would be saying that speed is the independent variable and time is the dependent variable. But the second sentence would compel us to consider speed as the dependent variable and time as the independent variable.

Although this chapter has focused on the graphical representation of mathematical equations, it has also discussed the nature of dependent and independent variables. We have seen that dependent and independent variables figure prominently in mathematics and in our world of everyday experiences even though the distinction between the two may sometimes be difficult to appreciate. But the distinction is one that is of great significance because it compels us to look at the patterns of cause and effect in mathematics and, indeed, the physical world as a whole.

part 4

PROBABILITY AND GAMES OF CHANCE

America is a land of sporting enthusiasts and it boasts an incomparable array of recreational activities including professional and college sports ranging from football and basketball and baseball to ice hockey and horse racing and auto racing. Millions of spectators watch these athletic competitions each year. But one of the biggest recreational activity in the United States is not found in football stadiums or basketball arenas or even baseball fields but instead in the gambling casinos and the convenience stores found in almost every state. We are of course talking about games of chance—whether they are the blackjack and roulette games found in every respectable casino or the state lotteries that can be played in tens of thousands of stores across the nation. And the mecca of gambling in America is the city of Las Vegas, Nevada; it boasts the world's greatest concentration of gambling casinos, hotels, and, for those incurable romantics, roadside wedding chapels. But it is the lure of gambling, in its myriad of forms, that brings millions of tourists and billions of dollars to Las Vegas each year. Paradoxically, the games of chance that have been responsible for the fantastic growth of Las Vegas over the past half century are also of special interest to mathematicians because these games can be analyzed using probability theory. Indeed, this is not a purely academic exercise as one must have a basic knowledge of probability theory to have a decent chance of holding one's own against the "house" over the long term.

This is not to say that you can read a few books on gambling techniques and games of chance and expect to win consistently in a casino. This outcome is rendered very unlikely by the fact that these are "games of chance" that are played in the casinos and that, as such, no one can expect to win them consistently. After all, winning in a casino is dependent on the roll of the dice or the draw of the card. The likelihood of consistent winning is rendered even more remote by the fact that the casino typically has a built-in edge in every game so that it can expect to be profitable over time—even though a few gamblers here and there may win fairly big jackpots. But the gambler who wishes to have any chance of holding his own against the casino must at least have some familiarity with probability theory. It may not make you a better gambler but it may help you to understand better why you should not give up your day job to pursue a career as a professional gambler.

PROBABILITY

All games of chance begin with the concept of probability, which, you may notice, is not very different from the word "probably." Indeed, when we say that we will probably do this or that we are saying in a fairly imprecise way that there is a certain likelihood that we will do a particular thing. Like "probably," the term "probability" is used to refer to the likelihood that a certain event will occur. We might say that the probability the dentist will determine that you need a root canal is 40 percent or that there is a 50 percent chance that it will rain today. How we calculate the likelihood of such an occurrence is based upon the ratio of the total number of occurrences of the desired event (such as the landing of a coin on the "heads" side) to the total number of possible events (the landing of a coin on either the heads or the tails side).

A coin toss provides a straightforward illustration of probability theory at work and can help sharpen our understanding of the subject. First, we need to begin with an explanation of the terminology used by mathematicians when they talk about probabilities. An event is said to have no certainty of occurring if it has a probability of 0. By comparison, an event that is certain to occur has a probability of 1. We know that if we toss a coin, there are only two possible outcomes—the coin will land on "heads" or "tails." No doubt a few mathematics students have pointed out that this statement may not be entirely true because a coin could conceivably land on its edge, thus making probability experts appear foolish. But as the edge outcome is a very unlikely event and one that is inherently unstable in that the coin will eventually fall to one side or the other, we can basically ignore it for the purposes of our explanation. Thus we have two possible events that can occur—the coin will land on heads or tails. As either event has a certain probability of occurring, we know that their probabilities will be greater than 0. But because we cannot be certain whether the heads or the tails will show up on every single toss, we also know that the probability will be less than 1 for each event. If we consider the problem a little more closely, then we can see that on a single toss there is only one of two possible outcomes that can occur. As a result, we have a 50 percent or ½ probability that the coin will land on heads and a 50 percent or ½ probability that it will land on tails.

But the principles of probability theory can be different than the actual outcomes. It is easy to see that the heads or tails side should have an equal likelihood of showing on a given toss. But if we were bored one afternoon and decided to toss a coin 50,000 times, it is unlikely that we would have exactly 25,000 heads and 25,000 tails at the end. Even though the theory tells us that we should have an equal number of heads and tails over any given number of tosses, we may find that we end up with a few dozen or even a few hundred more of one side than the other. But the more tosses we do, the more likely it is that any imbalance in the number of heads and tails will even out over time. Of course we might have to toss the coin 50,000,000 or so times before we were able to reach that point but then we would be subject to carpal tunnel syndrome.

Suppose that you do not really enjoy tossing coins. What other ways might you be able to illustrate some of the principles of probability theory? Many people like to throw dice or play cards in the casinos. As the typical dice has six sides, we know that we have a ⅙ chance of rolling any single side. Similarly, we have a 1/52 chance of pulling any single card out of a deck of cards or a ½ chance of pulling a black card out of that same deck. The law of averages says that the actual occurrences of particular events will approach the frequencies predicted by probability theory as the number of trials or events increases. This law does not say that the outcomes will somehow end up all even at a given point, as would be the case if we had 500,000 heads and 500,000 tails after 1 million tosses of a coin. Instead it is perfectly plausible that we could have 495,000 heads and 505,000 tails at that point. But the law of averages says that the ratio of heads to tails will approach a value of 1 with greater and greater numbers of tosses.

But we need to be careful when we talk about probabilities because some things that are expressed in terms of probabilities are not really based on mathematical calculations. Instead they are based upon assessments of current conditions and past events. We have already noted how mathematical probabilities govern the likelihood that you will draw a particular card. But it is a completely different thing to talk about the probability that a company's stock price will go up or down, that a horse will win or lose a race, or even that it will rain or shine. Our probability theory is much less useful here because these "psuedo-probabilities" are not really based on mathematics but instead on a number of factors—many of which may have little to do with mathematics and, as such, may not be readily quantifiable. In trying to determine whether a company's stock will appreciate in the near future, for example, we will consider many things such as the previous earnings of the company, its product line, its market dominance, the quality of its management, and the state of the economy as a whole. Indeed, it is more akin to speculation than mathematics to try to figure out the probable direction of its stock price, because there are so many factors to be taken into consideration. And even if we do a good job ascertaining the relevant issues, we may still find our predictions completely wrong. After all, we cannot say that the price of Widget Company stock has a 50 percent chance of rising and a 50 percent chance of falling because there are so many variables that may cause it to move in one direction or another. The fact that the Widget Company is the world's largest manufacturer of leather fashions would seem to be a positive aspect, but that might be offset by the decline in consumer demand in recent years for all types of animal skin clothing. It may be that peoples' tastes are also changing in that they are now more interested in wearing light cottons than skintight black leather—a trend that could have serious repercussion for the Widget Company's bottom line and the sale of its most popular products. A sinking economy might also hinder the prospects of the Widget Company because people who fear they will be laid off may be less interested in purchasing the comparatively costly clothing products it manufactures. In summary, there is no precise mathematical model that can quantify all of the things that can affect the price of Widget Company stock and offer a certain prediction as to the direction it will move.

Predicting the weather offers another example in which the lingo of probability theory is used to warn of the likelihood of rain, sleet, snow, hurricanes, typhoons, tornados, and any other form of inclement weather—even though such predictions are at best an imprecise science. The fact that Weatherman Bob predicts that there is a 75 percent chance of rain tomorrow does not mean that there is a 75 percent chance that it will rain all day in every area of the television market. Instead, it is more of a hedged bet in which Weatherman Bob is expecting that there is a pretty good chance that some area in his television market will receive at some time some rain on the following day. The fact that the probability of rain is declared to be 75 percent, or 85 percent, or 95 percent is largely irrelevant from a statistical point of view, because such predictions cannot be made in the context of probability theory. There is no single value as to the probability that we will have rain or will not have rain. Indeed, Weatherman Bob could offer the prediction every single day of "partly cloudy, chance of rain" and be completely accurate—at least in Florida. But his prediction would not be statistically verifiable. The weather, like the price of Widget Company stock, is the result of so many different variables that it is almost pointless to talk about the probabilities of a given outcome.

INDEPENDENT AND MUTUALLY EXCLUSIVE EVENTS

When we delve into the deep dark recesses of probability theory, we have to consider the ways in which we conduct probability experiments. If we flip a coin ten times, each toss is independent of every other toss. The fact that we might get heads ten times in a row—though unlikely—would not surprise a statistician because each toss of the coin is an independent event. Even though it is human nature to think that a coin would be overdue to turn up tails after it had turned up heads fifty times in a row, no statistician would be very sympathetic to such a view. He would strenuously point out that the coin has no memory and that each subsequent toss would not be affected by the prior tosses. But the fact remains that humans—not coins—are the only ones who fret about an abnormally long series of single-sided tosses. Probability theory does not prohibit twenty consecutive heads from occurring with a coin but it does enable us to calculate the remote likelihood of such an event: 1 in 524,288. This probability is derived by taking the probability that a given coin toss will result in "heads," which is $\frac{1}{2}$, and then multiplying that fraction by itself twenty times ($\frac{1}{2} \times \frac{1}{2} \times \frac{1}{2}$ and so on), which gives us the fraction $\frac{1}{524,288}$.

Mutually exclusive events are simply those events that cannot occur at the same time or in the same trial. In a single coin toss, the toss for heads is said to be mutually exclusive of the toss for tails. No matter how hard you may try, you cannot get both heads and tails on the same toss. Similarly, you cannot draw two different card faces such as a heart and a club on a single draw of one card (unless, of course, you are using a magician's deck). But when you are dealing with a mutually exclusive event such as drawing a heart and a club at one time, then the probability that you will draw *either* a heart or a club is the sum of the individual probabilities. We can calculate this probability in a snap: After all we have a $\frac{13}{52}$ chance of drawing a heart and a $\frac{13}{52}$ chance of drawing a club from a deck because each suit has 13 such cards in a 52 card deck. The likelihood that we will draw either of these two suits is $13 + \frac{13}{52}$ or $\frac{26}{52}$ or $\frac{1}{2}$. Unlike speculating in Widget Company stock or trying to predict the likelihood that a tornado will obliterate Weatherman Bob's broadcasting studio, we can take some comfort in knowing that the probabilities surrounding the occurrence of mutually exclusive events are set in the proverbial stone.

PERMUTATIONS, COMBINATIONS, AND ASSORTMENTS

People who want to win lots of money in games of chance know that their odds of winning are small. The selection of a winning ticket in a state lottery, for example, requires that the entrant pick a small group of numbers out of a much larger group of choices. Because few of us have clairvoyant powers strong enough to pick winning lottery numbers, we could try to calculate the odds of winning a lottery. But most of us do not approach state lotteries based upon a cold-hearted analysis of the probable odds that we will indeed walk away with the winning ticket and untold millions of dollars. The reason most of us do not do such a dispassionate analysis of the odds that we will win the lottery is because the odds are so small that we do not really want to think about them. Indeed, it is not uncommon for the odds of winning a state lottery in which six or so numbers are drawn from a bin of fifty or so possible choices to be less than one in twenty million. Clearly these are not such impressive odds that you could take one of your lottery tickets purchased from the local convenience store and use it as a down payment on a house.

If you think of a lottery as being a form of entertainment and wealth redistribution and nothing else, then you should be able to avoid any major disappointments. And it may be that Lady Luck will smile down upon you on a magical day if you select the winning numbers. But those of us who have never picked a winning lottery ticket might want to consider the way in which the probability of a winning lottery ticket is calculated.

Most state lotteries have a set of possible numbers from which a select few are drawn each week. The person who selects the winning numbers will win the grand prize, which in many state lotteries can exceed millions or even tens of millions of dollars. In general, the winning numbers are drawn at random and in no particular order. So if you live in a state in which you have to pick six out of forty-eight numbers, for example, you do not need to select the numbers in a particular order—such as the order in which the numbers are drawn. All you have to do is to match the six winning numbers and you will suddenly discover a brand new world of unlimited purchasing power.

In a six-number lottery, the calculation of the probability of winning will depend on the total number of possible six-number combinations. In most lotteries, the order of the numbers does not matter so the arrangement of numbers is known as a combination.

If the winning ticket had to match the order in which the numbers were selected, then the arrangement of numbers would be known as a permutation. And if you could use a given number more than once and the order of the numbers mattered, then your arrangement would be known as an assortment. Needless to say, the odds of winning a lottery with a permutation arrangement would be considerably poorer than the already remote odds of winning a lottery with a combination arrangement. No state lottery has tried the assortment approach in which the same number could be drawn two, three, four, five, or even six times as would be the case if you had a bin in which a single ball was selected six times but dropped back into the bin before the next selection was made. Assortment lotteries would undoubtedly aggravate the tempers of lottery players even further.

How do we calculate the total number of choices for combinations, permutations, and assortments? In the case of the combinations that are used in the typical lottery, we would simply multiply the number of possible choices for the first position (the first ball to be drawn) × the number of possible choices for the second position (the second ball to be drawn) × . . . × the number of possible choices for the last position (the last ball to be drawn). Because order does not matter when considering combinations, we can solve such problems fairly easily. But it may be useful to consider an illustrative example involving more modest numbers than those involved in a lottery.

Suppose that we have a peewee hockey team that consists of seven players, of whom any five can be put on the ice at a single time. As hockey coaches, we might wonder during a quiet moment how many five-child combinations can be put together from our seven-player team. The answer is fairly simply as we would use the following equation which involves factorials,[1] a concept we shall explain shortly: $7! / 5! \times (7 - 5)! = 7! / 5! \times 2! = 7 \times 6 \times 5 \times 4 \times 3 \times 2 \times 1 / 5 \times 4 \times 3 \times 2 \times 1 \times 2 = 7 \times 6 / 2 = 42 / 2 = 21$. So we can arrange twenty-one different five-player combinations of players for our team. As diehard win-at-all-costs coaches, however, it is very unlikely that we will ever actually use more than a few of those five-player combinations because some of these possible combinations will work much better than others.

The number of possible choices for combinations pales in comparison to those available when considering permutations. If we consider our hockey team, for example, we find that the number of permutations for selecting five-player groups from a seven-player roster when order is important is obtained by the following equation: $7! / 7!$ $- 5! = 7 \times 6 \times 5 \times 4 \times 3 \times 2 \times 1 / 2! = 7 \times 6 \times 5 \times 4 \times 3 = 2{,}520$. So we find that there are more than 100 times the number of possible permutations as there are combinations with our hockey team.

The enormity of the number of permutations can be appreciated if we suppose that we have 6 numbered balls that we want to arrange in all the possible combinations. We would then utilize what is called the $6!$ (pronounced six factorial), which means that we would carry out the following operation: $6 \times 5 \times 4 \times 3 \times 2 \times 1$ $= 720$. So these 6 balls could be arranged in a total of 720 possible ways. If we were really in a permutating mood, we could go to 12 balls and utilize a 12 factorial $(12!)$, which would be equal to $12 \times 11 \times 10 \times 9 \times 8 \times 7 \times 6 \times 5 \times 4 \times 3 \times 2 \times 1 = 479{,}001{,}600$. So you could spend quite a long time playing with your 12 balls and arranging them in all the possible ways.

Assortments offer a third method of counting in which a group of objects are selected from a larger group but can be selected more than once. Assortments can be very handy when you have to determine the number of telephone numbers that are available for a given geographical area or the number of possible letter and number combinations that can be used for a state's license plates. The formula for determining how many seven-digit telephone numbers can be created for a particular area is similar to that expressed above for combinations: the number of possible choices for the first position × the number of possible choices for the second position × . . . × the number of possible choices for the last position. In the case of telephone numbers, we know we can have 10 possible numbers for the first position (0–9), 10 possible choices for the second position (0–9), 10 possible choices for the third position (0–9), and so on, until reaching the seventh and final position. No doubt the mathematics students reading this passage will quickly conclude that the outcome is equal to $10 \times 10 \times 10 \times 10 \times 10 \times 10 \times 10 = 10,000,000$. So we know that we can have a total of 10,000,000 possible telephone numbers using 7 digits in our calling area. The sad fact of life is that people have such a high demand for phone numbers for their telephones, their fax machines, their internet connections, their pagers, and their cellular phones, that 10,000,000 different telephone numbers will supply the needs of a only a small fraction of that number of persons. The obvious answer is to simply add more digits to the traditional seven-digit telephone number because each additional digit will increase the quantity of available numbers by ten times. The problem with taking this approach is that many people believe that they should not be required to dial more than seven digits in any local number.

License plates provide another example in which assortments are used. Most license plates are combinations of letters and numbers and, because of this mix, can provide us with the opportunity to express our literary talents by sneaking bad words onto the plates. How many possible arrangements of letters and numbers can be created for motorists in a state in which 6 letters and numbers may be used on any given plate? Because there are 26 possible letters (A–Z) and 10 possible numbers (0–9) that can be used in each space on the plate, then the typical license plate will have $26 \times 26 \times 26 \times 10 \times 10 \times 10 = 17,576,000$ possible arrangements of letters and numbers. This is very good news for most states because there are only two or three states that could have anywhere near as many as 17 million licensed vehicles on the road. As with telephone numbers, any perceived shortages can be addressed with the addition of an extra letter or number.

PASCAL'S TRIANGLE

Those who seek to understand the theory of probability should take a moment to learn about Pascal's triangle. This seemingly innocuous arrangement of numbers provides profound insights into the theory of probability. The discoverer of Pascal's triangle, Blaise Pascal, a seventeenth-century French philosopher and mathematician, stumbled upon an interesting arrangement of numbers that could be organized in a triangular pattern as follows:

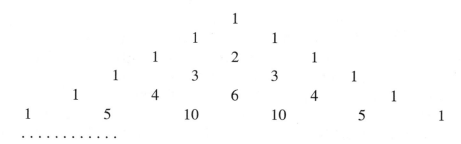

The pattern is fairly straightforward in that "ones" form the edges of the triangle and each number in each successive row is formed from the addition of the two numbers flanking it on either side of the preceding row. As a result, we find that the 2 in the third row is formed by the addition of the two ones in the previous row. Similarly, the 6 in the fourth row is formed by the addition of the two threes in the third row. This process can be continued indefinitely but the process of replications remains the same as new numbers are continuously created from the addition of flanking numbers in the previous rows.[2]

What does Pascal's triangle have to do with probability theory? It can be used to illustrate the likelihood that a specific number of coin tosses, such as 6, will result in 1, 2, 3, 4, 5, or 6 heads. Now this might be of some use to us at the office if we get involved in a bet as to the number of heads that might turn up in a series of tosses. Even though such a contest might itself have a very low probability of taking place, we have to admit that Pascal's creation provides a wonderful summary of the possible outcomes in such multistep events as illustrated by the following example: The sixth row of our triangle has the numbers, 1, 5, 10, 10, 5, and 1 when read from left to right. But one might quickly point out that the numbers are the same when read from right to left because these probabilities form a symmetrical pattern. The way to interpret this row of Pascal's triangle is as follows: There are a total of 32 possible outcomes. The first outcome—one head and five tails—has a $\frac{1}{32}$ chance of occurring as does the last outcome—five heads and one tail. The second outcome—two heads and four tails—have a $\frac{5}{32}$ chance of occurring as does the second to last outcome—four heads and two tails. The third outcome—three heads and three tails—have a $\frac{10}{32}$ chance of occurring as does the fourth outcome—three tails and three heads—or a total of $\frac{10}{32}$ chance. All of these possible combinations add up to $\frac{32}{32}$ or 1 and thus make up all of the possible outcomes of tossing a coin 6 times.

What makes Pascal's triangle particularly interesting, as pointed out by the American author Brian Burrell in his *Guide to Everyday Math*, is that it shows how events that we would expect to be the likeliest to occur do not necessarily occur as often as we might expect: "If you consider the $n = 6$ row of Pascal's triangle, it says that the probability of getting 3 heads in a 6-coin toss is $\frac{20}{64}$, which is less than a third."[3] This means that a little less than one out of every three tosses will generate the same number of heads and tails. This seems very strange to us because we would think, based upon the thundering pronouncements of the statistics gods, that we would find it very difficult not to toss an equal number of heads and tails with our 6 tosses. But Pascal's triangle tells us otherwise. The important point is that even though the three-head/three-tail outcome is the most common outcome, it is far less likely to occur than all of the other possible outcomes combined.

CARDS AND DICE

Many people enjoy playing a game of cards or tossing a pair of dice as a source of entertainment. Indeed, some people view these games as a very civilized way to compete with each other. But all games of chance must ultimately answer to the laws of probability. Those persons who wish to improve their chances of at least holding their own against their fellow card players should at least familiarize themselves with the mathematical odds of drawing particular cards or hands.

Most of us play with a deck having fifty-two cards, consisting of four suits of diamonds, hearts, spades, and clubs. Each suit has thirteen cards consisting of numbered cards from 2 to 10 and the "face" cards—the jack, queen, king, and ace. Probability theory and cards are inextricably intertwined with each other because card games such as blackjack and poker are among the most common forms of gambling that would-be millionaires encounter when heading into the casinos. It is virtually impossible to prevail against the casino in the long term because it does not pay out all of the wagered moneys to the winning players. Instead it retains an amount equal to at least several cents on the dollar for each bet that is made. To overcome this inherent disadvantage, serious card players try to familiarize themselves with the odds that certain cards will turn up as they progress through their game. We know that a standard card deck has fifty-two cards so that we have a $\frac{1}{52}$ chance of drawing any single card from a deck. But the odds of drawing any card will change if we do not replace the drawn cards. If we draw a card such as a king from the deck, then we will be left with fifty-one cards. As there are only three kings left, the odds of drawing another king is $\frac{3}{51}$. Assuming that the second card drawn is indeed another king, then the odds of drawing a third king is $\frac{2}{50}$ because there are only two kings left in a deck of fifty cards. So we would calculate the probability of drawing three kings in a row by multiplying $\frac{1}{52} \times \frac{3}{51} \times 250$ $= \frac{24}{132600} = 0.00018$—which are fairly slim odds indeed.

People who "count" cards try to improve their odds of winning by keeping track of the cards that have been drawn so that they have a better idea of the likelihood that any particular remaining card will be drawn. If one knows that most of the high cards have been drawn from a deck, then one can adjust one's betting strategy accordingly. Casinos do not like card counters for the very reason that they tend to be more successful in their betting than the average run-of-the-mill gambler. To thwart card counters, casino operators have resorted to other means, including using multiple decks of playing cards, rotating dealers from table to table, and, in some extreme cases, refusing card counters entry into the casino itself.

Poker is one of the most popular card games. In draw poker, each player is dealt five cards face down. Each player may discard up to three cards and draw replacement cards from the deck. After all the players have completed their draws, they will then begin to bet and try to bluff each other. Poker is a game in which there is a hierarchy of card combinations beginning with the royal flush (consisting of five cards, the ace, king, queen, jack, and 10, of the same suit) which has a probability of 0.000001539 and occurs about once in ever 650,000 hands. At the other end of the spectrum is the lowly pair of cards (two 2s, two 3s, two 4s, etc.), which has a probability of 0.42 and occurs once in every 2.4 hands. Because of the immense unlikelihood of being dealt a royal flush even a single time, you would find it very suspicious if your opponent was dealt a royal flush two, three, or even four consecutive times, using a single deck of cards. Indeed, the probability of drawing three successive royal flushes would be so remote as to compel you to leave the table. If you were to try counting cards, you would find yourself trying to keep track of the cards drawn in each hand of poker so as to have a better idea as to when the bluffs of the opposing players would be more believable and, hence, when you should increase or decrease your own bets.

Blackjack or 21 offers another game in which card counting may be helpful. Each player is dealt one card face down and the second card face up. The object is to get as close to 21 points as possible without going over; a player can take as many "hits" (cards) as he wishes to reach that end. Although there are different strategies used by blackjack players, the most common approach is to take additional hits until one has at least 17 points. Anything up to 16 points will usually be regarded as low enough to warrant another hit. However, this strategy may vary depending on several factors, including the cards showing in the dealer's hand, the cards showing in your own hand, and the extent to which certain "high" cards have already been dealt in the game. You also have to remember that the house will win if you and the dealer have the same number of points in your hand.

Many games of chance also use dice. As with a deck of cards, it is equally likely that any of the six sides of a dice will show up on a toss. Each side thus has a ⅙ chance of turning up. There is a phrase that cards and dice have no memory so there is nothing to prevent the same side of a dice from turning up over and over again. However, we human beings cannot help believing that a dice that has rolled four consecutive "fives," for example, is overdue to roll a different number because we will feel—regardless of what the statisticians tell us—that the luck of the fives must run out. As a result, we may alter our betting based upon our belief that the dice face will be different on the next roll even though there is no statistical reason why a five would be any less likely to show up than any other side of the dice.

In games in which two dice are used at the same time, the odds can be calculated for two six-sided dice where the total number of spots range from two (each dice has one spot) to twelve (each side has six spots). Because each side of each dice has a 1/6 possibility of turning up, there are a range of probabilities for the total number of spots between two and twelve. Indeed, the 2 and 12 spots both have a ⅟₃₆ probability of turning up, whereas the 3 and 11 spots have a ⅖₃₆ probability of showing, and the 4 and 9 spots have a ⅗₃₆ probability of landing face up, and so on. Ultimately, there are 36 possible combinations that can be rolled with the two dice. All of the possible combinations summed together will have probabilities equal to $^{36}\!/_{36}$ or 1—which means that all of the possible choices have been exhausted and that one of the combinations must necessarily occur on any toss of the dice.

All persons familiar with probabilities know that gambling is not the royal road to riches unless they are the proprietors of gambling casinos. All gambling is fundamentally related to probability theory even though it is rare to overhear a conversation at a blackjack table about mutually exclusive events or Pascal's triangle. But it is the theory of probability that provides would-be gamblers with some insight as to the likelihood that they will win a massive jackpot at the tables or go home poorer.

eleven

Very few mathematics students would quarrel with the idea that the everyday world of human affairs is a cacophonic spectacle of sights and sounds. We live on a planet that is increasingly crowded by its growing population of several billion souls, most of whom can be found talking about the affairs of the day on their cellular phones. The United States itself also boasts a population of several hundred million who engage in a myriad of activities ranging from careers and sporting events to education and religious worship.

If a nation is to have a huge population, then it must have some idea as to the basic features of that population. In other words, we need to know something about the numbers of persons who are born, who are alive, and who die, because it gives us a better idea as to the amounts and allocations of resources that will be required by the population for it to survive and flourish. Indeed, the government carries out a census every ten years to ascertain certain basic characteristics about the citizenry such as sex, household size, income, and so forth. The government, of course, has a vital interest in knowing about its population because it is charged with the task of providing certain essential services for that population, including national defense and various forms of governmental assistance. Unfortunately, the government must also pay for these very same services and so it must impose taxes upon these very same citizens. If the government had no clue as to the size or basic characteristics of its population, however, then it would have no idea about the demands that would need to be met by its services.

Yet the government is only part of the story because it does not own the means of production in the country nor does it provide the bulk of the income for most of the population. This task is left to private enterprise where the ingenuity of humanity is able to create jobs that in turn generate incomes that in turn can be used to pay for goods and services that ultimately keep the economy humming. Here, it is vital for private businesses to know something about the characteristics of the population as a whole so that they may determine the appropriate amounts of goods and services to produce as well as the number of employees they need to hire to produce those goods and services. But it is up to the government to count these people and try to offer general conclusions about their behavior. However, it often falls to social scientists—particularly statisticians—to glean meaningful information from these figures, typically through the use of statistical studies.

Statistics is not a field that engenders great enthusiasm among the general public because most people associate it with dreary tables of numbers and diagrams. After all, it is not very interesting to many people to talk about the number of births or deaths in a given year—unless, of course, you are the one who has managed to outlive all of your rivals and competitors. In any event, the importance of statistics to government and industry alike stems from the unavoidable fact that we must know something about the population as a whole in order to try to both manage the affairs of that population and provide for its needs.

Statistics was invented by an English gentleman named John Gaunt who believed that many social ills that existed in sixteenth-century England could not be profitably attacked unless society knew something about the basic characteristics of its population. Gaunt himself was a haberdasher who lived at a time in which public health was a growing topic of concern with the population being decimated periodically by outbreaks of the plague and other infectious diseases. England in the seventeenth century was not exactly the model of public sanitation as the streets served as makeshift lavatories and the public water supply was not something that one would necessarily want to bottle. Gaunt himself was some-thing of a social activist who believed that accurate statistics about England's population would make it possible to improve social welfare or, at the very least, keep better track of the corpses. In any event, Gaunt himself became deeply involved in studying the birth and death rates of people in various English cities. This desire to better understand the basic characteristics of the English population became Gaunt's lifelong passion.

One can only imagine the extent to which Gaunt's dinner companions must have appreciated his tales of death and misery. But one would have been struck by Gaunt's discovery that the rates at which people died from natural causes, suicide, accidents, or homicides did not change very much over time. This finding surprised Gaunt be-cause he had not really considered that the seemingly chaotic demographics of the citizenry to have any sort of regularity or predictability at all. Gaunt's studies also revealed that there were slightly more males than females born each year but that males—due to their proclivity to engage in dangerous occupations and go to war—died at greater rates at almost every age level. This finding may have knocked some wind out of the idea of women being the "weaker sex" but it showed, at the very least, that one could paint a picture of the population at a given point in time and draw valuable insights from information relating to its vital statistics. Gaunt's work was encouraged by Sir William Petty, a professor of anatomy, who realized that Gaunt's work could have far-reaching implications for assisting the authorities in planning for the needs of the population. Petty was also very adept with numbers and believed that Gaunt's work offered a promising beginning for quantifying all types of human behavior. As such, Petty believed that the social sciences could be put on as precise a mathematical footing as the physical sciences. Undoubtedly, Petty was inspired to this goal by the great advances that were being made in physics and mathematics at the time, particularly with the development of the calculus and Newtonian physics. Gaunt's conclusions regarding the recurring behavioral patterns that could be found in large groups of human beings inspired Petty to dub the embryonic science of statistics as "political arithmetic" and write influential essays on the topic.

But Petty was not exactly on the "A" list of influential people in seventeenth-century England; his ideas regarding the use of statistics to describe gross features of the population were disseminated throughout the land but were ignored. Indeed, the field of statistics was something of an intellectual graveyard for more than a century, viewed as more of a curiosity than a field in which laws to predict human behavior could be gleaned. It did not bear a great deal of resemblance to the subject that we now study in school because statistical analysis at that time (if we can call it that) consisted primarily of collecting raw data and drawing conclusions about the central features of that data such as the mortality rates. Gaunt and Petty, despite their wonderful contributions to the social sciences, had not really developed the sophisticated analytical techniques that would later make modern statistics so indispensable to the government and industry alike.

MEANS, MEDIANS, AND MODES

Having wandered through the ancient history of statistics, we need to consider the basic concepts of statistics so that we can acquire a cursory knowledge of the subject. Perhaps the best place to begin is to familiarize ourselves with certain mathematical devices for obtaining information about a given population—the mean, median, and mode. We can imagine an office called Smith and Associates with seven employees who make the following salaries each week: $100, $500, $500, $600, $800, $1,000, and $1,400, respectively. Not surprisingly, the president and founder of the company, J. L. Smith, receives the highest salary. Smith's mantra that all things arise from hard work pervades the entire office and is a point of inspiration to all the employees, particulary his nephew, Bob Smith, who recently began his employment as senior vice president with a salary of $1,000 per week. The rest of the staff salaries trend downward all the way to the lowest-paying position, the washroom attendant.

The mean itself is perhaps the most widely known statistical measure and is calculated by simply adding up all the quantitative values being considered (the gross salaries of the persons working in the office) and then dividing that total by the number of employees (7). So we would determine that the total weekly salary (the office payroll) is equal to $4,900. The average salary of the persons working in that office is equal to $700. Yet the average says nothing about any one individual's salary. The mean is also not very helpful in providing information about the distribution of salaries in that office because it can be overly skewed upward or downward by a particularly large or small salary. In the case of our office, the salaries being paid to the two Smiths inflate the average; if the two Smith salaries were not included, then the total weekly payroll in the office would be $2,500, and the mean salary would be $500.

The median, by contrast, provides us with some clues about the salaries in the office but it is also incomplete. It is the middle value for which there are an equal number of salaries both above it and below it. In the case of the Smith office, the $600 salary would be the median because there are three smaller salaries and three bigger salaries. But because the median value does not really incorporate the outlying values of the group (the biggest and smallest salaries), it cannot really tell us much more than the mean about the distribution of the salaries. As with the mean, the Smiths draw salaries that are far in excess of the other employees. As a result, the median value does not really say anything one way or another about the size of the biggest or smallest salaries.

This conclusion leads us to our third measure—the mode, which is a measure of the value (salary) that occurs most frequently in our array. In the case of the Smith company, the mode is $500. This is not overly helpful because only one salary—that of the washroom attendant—is smaller; the other four salaries are greater than the mode. So this entire exercise leaves us to wonder whether we have accomplished anything of lasting significance in calculating the mean, median, and mode of the salaries at the firm. Each measure provides some clues about some of the salaries at the firm but none of these measures is able to tell us about the range of distribution of the salaries.

Some people might be content to simply shrug their shoulders and say how much they enjoyed the discussion about means, medians, and modes. But they should take heart because statisticians know how to measure the distribution of values in a given case. Indeed, they even have a special name for this concept—the standard deviation, which provides a very useful measure of the extent to which values of any given set of data cluster around the mean. And it is quite gratifying to learn that the method for calculating the standard deviation can be done very easily. Statisticians use the Greek letter sigma (σ) to represent the standard deviation so that they do not have to write out the words "standard deviation" repeatedly (which can be very burdensome when you have spent several hours, pen and paper in hand, trying to solve mathematical problems). The important thing to remember is that the standard deviation is first and foremost a measure of the distribution of the values in a given set of data. And anyone can follow the simple six-step process needed to determine the standard deviation for any set of data.

So how do you calculate standard deviations? It is very simple.

First, you have to have a set of numbers or values that make up the set of data points that you are trying to measure. For our purposes, the set shall consist of the numbers 1, 3, 5, 7, 9. To make this a little easier to follow, we shall imagine that these five numbers represent the number of dogs owned by five neighbors living in a community. Second, we need to determine the mean of this set of numbers, which is equal to $25 \div 5 = 5$. So the average number of dogs owned by any single neighbor is 5 dogs. Third, we would need to find the difference between the mean (5) and each of the numbers 1, 3, 5, 7, 9 (the number of dogs actually owned by each of the neighbors), which gives us 4, 2, 0, 2, 4. Fourth, we would then square these five numbers and add them together $(16 + 4 + 0 + 4 + 16)$ to get 40. Fifth, we would find the mean value of these squares $(40 \div 5 = 8)$. Finally, we would calculate the square root of this mean and determine that it is equal to 2.82. So we are left with a value for the standard deviation equal to 2.82 as compared to the original mean (average number of dogs owned by each neighbor), which is equal to 5. In short, this standard deviation tells us the extent to which the square root of the mean of the squares of the individual deviations varies from the mean value of the set of data points (the number of dogs). The value for the mean of the five data points differs significantly from the standard deviation which in turn tells us that the data points are distributed fairly far apart. In other words, the actual number of dogs owned by each neighbor varies considerably from the mean or average (5) that we calculated for the entire neighborhood.

The other important feature of standard deviations in normally distributed sets of data will be characterized by the following pattern: 68 percent of all values will fall within one standard deviation of the mean, 95 percent of all values will fall within two standard deviations of the mean, and, last but not least, 99 percent of all values will fall within three standard deviations of the mean. So this will provide us with very useful information as to the distribution of all of the data points around the mean. But it is also true that not every collection of data points will be distributed in a nice neat bell-shaped (normal) curve around the mean. But there are very good mathematical techniques for calculating standard deviations when we are not fortunate enough to have a normal frequency distribution.

No doubt you are wondering whether we are accomplishing anything by discussing the way in which standard deviations are calculated. But such a discussion is very helpful from the standpoint of showing us in an abbreviated manner the way in which all of the values are spread out. This in turn will affect the degree of confidence we have in that data. But the distribution of data points —regardless of whether they represent the number of dogs owned by persons or the age of those persons or the number of ex-spouses of each of those persons—can be very useful to those who wish to sell products to the persons (or their dogs). This information can also be useful to other entities such as the local government because it may want to identify the number of persons and dogs living in the neighborhood.

Statisticians often use the same two-dimensional graphs with both an x-axis (horizontal) and y-axis (vertical) that we have discussed previously. The two axes are parallel to each other and intersect at a point of origin usually described with a "0." The frequency of the observations or events or data points are usually marked off on a scale on the vertical axis. The horizontal axis is typically used to describe the matter being considered such as the weight of the typical male in a town or the make of car most frequently involved in highway fatalities. But the collection of this data can be arduous because it can involved making literally hundreds or even thousands of entries into our record books.

Someone once said that a picture is worth a thousand words. Nowhere is the truth of this adage more apparent than when the seemingly randomly distributed values of several hundred data points are plotted on a graph, thereby making it possible to draw certain conclusions from the general pattern formed by these points. Suppose we want to know something about the distribution of thin people and obese people in our community. One way of obtaining that information would be to head out into the community and go door to door with a portable scale and ask people to weigh themselves. We might try to persuade them of the importance of our work and reiterate our concern over the poor diets of most Americans as well as our belief that proper nutrition is the key to remedying this serious health problem. We would then weigh each of our participants and record the information and, as we made our way through the neighborhoods, begin to see some weights appearing more frequently than others in our charts.

We should point out that the success of any statistical study necessarily depends on the extent to which we can select a sample that is representative of the population as a whole. Of course if we were very ambitious statisticians, we might decide to try to weigh every person in the city or the state or even the country. But we might find our staff of researchers overwhelmed with having to get thousands or even millions of persons to step onto our trusty scales. So we would want to use the sampling techniques that would enable us to select a comparatively small group of persons whom we believe would be representative of the population as a whole. Of course if we were interested only in the weights of those people whom we were visiting, then we would not be concerned about the reliability of our sampling process: We would be weighing every member of the particular group that was to be featured in our study so our sample and our population would be equal. But we would have to know something about selecting a representative sample group if we have any desire to extrapolate our findings to the greater population as a whole.

Suppose that we went out and weighed all of the individuals in our sample and recorded all our results in our notebooks. We would then want to consider how we could summarize the data in a form that would be both comprehensive and easily comprehended. Even though we might be very comfortable with constructing extensive tables listing all of our measurements, we might find it very cumbersome to review these compilations repeatedly to try to keep track of how many men weigh 150 pounds or 170 pounds or 250 pounds. We could summarize the results in a more concise table that would have ranges of weights (e.g., 150–159 pounds, 160–169 pounds, 170–179 pounds, and so on) along one side and the number of persons that fell within each of these categories next to that particular category. So if there were 17 men who weighed between 150 to 159 pounds, we would write "17" next to that interval. If there were 21 men who weighed between 160 to 169 pounds, we would write "21" next to that interval, and so on. Our staff would eventually create a very nice summary of all of the weights that we were able to obtain from the citizens of the town. We would then have a good summary of the distribution of weights in our group of participants. But because we are visually oriented, we might find it more useful to provide a graphical description of our data. Perhaps the simplest approach would be to return to our x-y axes and plot the weight intervals on the x-axis and the number of persons appearing within each of those intervals on the y-axis (see figure 1).

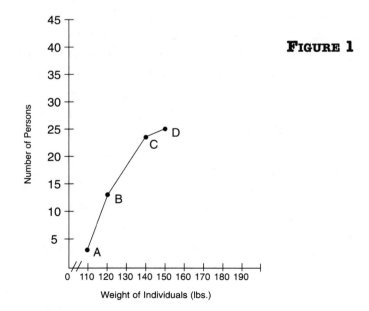

FIGURE 1

Although we could begin with the smallest possible interval, which would be 0 to 10 pounds, we would not find grown men who would fit within these parameters. Thus, we might want to start our weight intervals with the least massive participants in our study who, we might find, tip the scales at 115 pounds. As a result, we could begin our weight intervals at 110–119 pounds and then move up the appropriate amount on the y-axis and plot the number of persons falling within this interval. If there were only these three persons, then we would place a point as indicated at A on figure 1. This point would represent the intersection of one line projecting outward from the "3" mark on the y-axis and a second line projecting upward from the middle of the 110–119 interval on the x-axis.

We would then move to the next interval, the 120–129 pound group, and repeat the process, plotting the point (point B) that would correspond with the 12 persons that fall within this interval. The 130–139 pound interval might have 23 persons (point C), the 140–149 pound intervals 25 persons (point D), and so on. But we would probably notice that at some point the number of persons being included in a given interval would stop increasing and instead begin decreasing. In other words, we might find that after plotting an all-time high of 46 persons in the 180–189 pound interval (point H), we would then tally only 34 persons in the 190–199 pound interval (point I), 24 persons in the 200–209 interval (point J), and so forth. After we finished plotting all of the points of all of the intervals, we would probably find that the pattern of our sample resembled a familiar shape, the so-called bell curve (see figure 2).

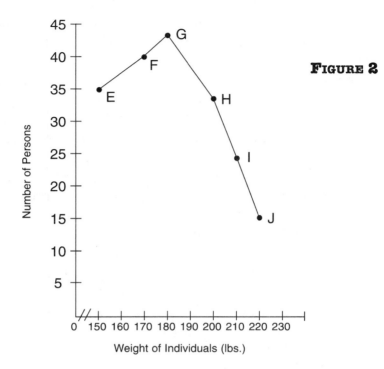

FIGURE 2

What is the significance of this pattern? The fact that the plotted line begins at a fairly low level and then rises to a point before falling back to another low level suggests that there is a frequency distribution that is strikingly symmetric and, dare we say, predictable. This is not to say that this curve is perfectly proportioned. Yet given an adequate number of observations (or, in this case, participants who were not afraid to step onto a scale), it will more and more closely approximate a symmetrical bell-shaped curve. It is very helpful to see this chart because it encompasses in a very simple and direct way all of the information that we had previously collected and tabulated. We can thus look at this curve and determine the number of people who fell within each of the intervals and perhaps draw some conclusions about the prevalence of thin people and obese people in the community. But we also see that the highest point of the curve above the x-axis is the interval in which the most participants in our survey were registered. As we move to either end of the curve, however, the curve itself approaches the x-axis, thus indicating that successively fewer and fewer participants are in each of those intervals. Indeed, we would expect it to be the case that the most common weight of the men in our sample would cluster around some normal or average standard and that there would be progressively fewer persons appearing in the outlying intervals.

We would not be surprised to find that fairly few of our participants (8) weigh between 260–269 pounds because we do not meet many people in our daily travels who weigh more than 260 pounds. That being said, we would find it very strange if an even more extreme interval—those between 370–379 pounds—boasted 12 people. This is an even more uncommon weight and so we would wonder why there was such a comparatively large (no pun intended) number of people in this group. This finding might force us to reconsider our sampling techniques and see whether our sample was truly representative of the population we were surveying as a whole. Our review might show that we had inadvertently stumbled onto a sumo wrestling camp in one of the neighborhoods and included all of the participants in our study. The problem, of course, is that there are not really very many people who engage in sumo wrestling in the population as a whole. If, for example, there are only 50 professional sumo wrestlers who engage in sumo wrestling in the United States as a whole, then they are obviously a very tiny portion of the population (approximately 280 million, or 0.000017 percent). But if we found that our study of 400 people had included 35 of these 50 sumo wrestlers due to their having attended a self-esteem seminar in one of our canvassed neighborhoods at the time of our study, then the sumo wrestlers would have constituted 8.75 percent of all of the persons in the study—an extremely disproportionate representation of sumo wrestlers as compared to the population at large.

Assuming that we are able to derive a sample of participants that is representative of the population as a whole, however, we are going to find that the plotted points will approximate the curve known as the normal frequency curve. Now whether you consider this curve to be normal in any sense of the word is really a matter of taste, but it refers to the aforementioned bell-shaped curve that is perfectly symmetrical. In other words, the two halves of the curve are mirror images of each other. This ideal representation is rare when dealing with fairly few data points; one may have to plot several hundred points to obtain a perfectly symmetric bell curve. But having acquired many data points does not necessarily guarantee a curve that would mimic a normal frequency curve. After all, we may simply not have the requisite data points that would fit within a perfectly symmetrical curve. But there is a rule of thumb that the greater the number of observations or data points, the more closely the curve will mirror that of a normal frequency curve.

CONFIDENCE LEVELS

Statisticians are also concerned about the reliability of the conclusions that may be drawn from their surveys. Suppose I am a manufacturer of the Zippy Cola product line and decide to branch out into exotic-flavored carbonated beverages. Having not just fallen off of the proverbial turnip truck, I would want to test my new products on selected samplings of individuals in order to gauge their reaction to my offerings and see if these products could have a viable commercial future. So I might have my marketing department gather together several samples of prospective tasters who would hopefully be representative of the population as a whole—or at least the population that I believe will buy my new products.

After the taste tests are complete, I would be very pleasantly surprised to find that 85 percent of the respondents said they would buy my carbonated celery beverage and an astounding 97 percent of the respondents would buy my peanut cola. These are the type of responses that all manufacturers want to see for their products, but

I would have to be careful not to get too carried away. After all, what assurances do I have that these responses are accurate and that I can be confident in retooling my factory to produce millions of gallons of carbonated celery and peanut cola each year?

My statistician might claim that our survey is accurate to within two percentage points. Of course I would want something more substantial in the way of assurances, to which he might respond that our beverage samplers were representative of our targeted audience. Nevertheless, I might still have some trepidations about whether our sample groups totaling 1,000 persons could speak for the entire nation of nearly 300 million people. Here is where knowing something about confidence intervals can be very useful for evaluating statistical studies. A confidence interval is based on the standard deviation and would be approximated using the following formula: The standard deviation is equal to the square root of the product of the number of persons who said they would buy Zippy peanut cola multiplied by the percentage of persons who said they would not purchase Zippy peanut cola. In the case of our sample the standard deviation would be equal to $\sqrt{850 \times 0.15} = 11.29$. Now if our responses approximate a normal frequency distribution, then we can be confident that there is a 95 percent that an outcome falls within two standard deviations of the mean. In this case, the standard deviation is equal to a little more than 1 percent of the sample (11.29 divided by 1,000), so two standard deviations is equal to 2 percent. In this way, we can say that our sample gives us a confidence level of 95 percent that our survey is accurate to within a degree of error of 2 percent.

CORRELATION

When choosing to carry out statistical studies, one must necessarily deal with the real world of messy data and inexact results. Statistics itself is often used to study human behavior even though it has a variety of other applications. But predicting the behavior of human beings as individuals is notoriously difficult. If I were the owner of a flower shop, I might expect that my loyal clientele would respond to an across-the-board 10 percent cut in prices by making even more purchases of flowers than usual. But even the best statistician in the world would not be able to tell me which of my customers would respond to the sale and buy flowers and which of my customers would ignore my advertisements. Although I would reasonably expect that such a grand sale would motivate many of my current customers and perhaps a few new customers to come into my store, I would have no way of knowing which particular customers would respond beforehand. Indeed, I would not really be so concerned about knowing which of my customers would show up so long as I knew that some percentage of my customer base would come banging on my door. But I would be very concerned that the increased volume would more than make up for the 10 percent discount.

When statisticians talk about correlation, they are referring to the extent to which there is a statistical relationship between two variables. In other words, the fact that I decrease my store prices and it results in increased sales might be an example of a correlation between these two variables—lower prices and increased sales. But statistics is not a precise subject due to the vagaries of the subject matter with which it must deal. So we can never be sure whether there is a precise causal relationship involved between lower prices and increased sales. Yet most of us would intuitively accept the idea that a lowering of the prices for flowers would bring hordes of people to my store. So the concept of correlation is essentially one of changes in one variable being associated in some way with changes in another variable. Statisticians themselves measure the amount of correlation between two variables on a scale of 1 to −1. A correlation with a measure of 0 would indicate no relationship at all between the two variables being considered. A measure of 1 would register in situations in which the second variable invariably changes in the same way as the first. A measure of −1, by contrast, would occur if the second variable always behaved in a completely opposite way from that of the first. The closer we move to a correlation of 1 or −1, the stronger is the relationship between the variables; a movement from either extreme toward 0 indicates a correspondingly weaker correlation. To return to our flower shop example, we might expect a correlation nearer to −1 if attempting to relate our drop in prices with increased sales of our roses. Conversely, we might find that there was a direct correlation between the working hours of the staff and their blood pressure levels and that the measure of correlation would be closer to 1.

When statisticians plot out all of their data points on a two-dimensional graph, they use mathematical techniques to derive the regression line; this is a straight line that minimizes the differences among all of the points on the graph. It thus offers the truest description of the functional relationship between the two variables at any given interval even though it may not precisely mirror the plotted values of the functional relationship. But it is an important graphical tool for illustrating the ways in which one value changes with respect to another.

Statisticians today have far more sophisticated quantitative techniques than existed in seventeenth-centiury England when John Gaunt first began to study the birth and death rates of his fellow countrymen. But Gaunt's spirit remains with today's statisticians because they are driven by the same desire to bring order and organization to the often chaotic landscape of the physical world. Contemporary statisticians are able to use sophisticated mathematics to evaluate their data in a variety of ways that could not have been envisioned by Gaunt who lived at a time when mathematics was only beginning to be used to clothe the physical sciences with a vestige of respectability and legitimacy. Indeed it is striking that the melding of mathematics and science and the creation of what we now know as the scientific method took place in a comparatively short time in Renaissance Europe. It is all the more remark-

able that this staggering intellectual upheaval was due to the courageous and extraordinary works of a very few courageous men—
Copernicus, Kepler, Descartes, Galileo, and Newton—whose lives
and works will be explored in the next few chapters.

part 5

MELDING MATHEMATICS AND SCIENCE

twelve

Mathematics can be enjoyed as a purely intellectual exercise or it can also be used as a powerful tool for solving complicated problems. Much of this book has focused on this later aspect of mathematics because it has arguably been the single most important tool for helping humanity understand the perplexing complexity of the universe. Mathematics has enabled us to conceptualize the dynamic processes of the universe and describe widely disparate phenomena, ranging from the motions of the planets to the paths of falling objects, using the same kinds of numerical representations. In this way, mathematics has provided a method for quantifying an incredible array of physical processes—regardless of whether they occur in our world or on the farthest stars. The previous chapters have highlighted just a few of the ways in which mathematics has been used to bring about a greater under-standing of and more profound insights into the nature of our own world as well as the human condition itself.

But the use of mathematics as a means for studying both the

world around us as well as the vagaries of human behavior is a comparatively recent development. Indeed, mathematics existed largely separate and apart from what we now call "science" for much of human history.[1] Although mathematics had offered both intellectual and practical rewards for thousands of years—being used for purely mundane transactions as well as being studied in a deeper philosophical sense by the Greeks, particularly Pythagoras —it had not really been used in conjunction with scientific investigations in any meaningful sense until the sixteenth century. One of the reasons for this was the embryonic state of the sciences themselves. In sixteenth-century Europe, for example, there was more interest in alchemy and astrology than in what we would now call chemistry and astronomy. Certainly a good astrologer was considered at that time to be far more valuable than any astronomer. Moreover, there was no widely recognized form of scientific inquiry akin to our modern scientific method whereby the investigator forms a theory, collects evidence, tests the theory, and draws conclusions. The world was not viewed in terms of basic physical properties and underlying forces. Rather, it was perceived as a mysterious realm of spells, demons, witches, and monsters.

The wrenching changes brought about by the fusing of mathematics with the physical sciences coupled with the creation of the scientific method were stupendous in scope. The primary focus of creative inquiry changed. It evolved from one of seeking to validate existing philosophical and religious paradigms (such as the church-sanctioned geocentric model) to a more open-minded approach. The latter involved systematic attempts to observe and understand the surrounding world on its own terms. In short, investigators were no longer consumed with trying to shore up a theologically inspired model of the cosmos purely for its own sake. They wanted instead to learn about the actual workings of the universe.

The unification of mathematics and science was a long and arduous process that occurred over the course of several centuries, beginning in the sixteenth century with the work of Nicolaus

Copernicus, who offered a sun-centered theory of the universe, and, later, with the discovery by Johannes Kepler of his three laws of planetary motion.[2] Kepler was particularly critical in the development of a mathematically based science because he was the first to formulate mathematical laws that accurately described the orbits of celestial objects around the sun. Thus, mathematics could be used to predict certain phenomena like celestial orbits as long as one had the correct equations and enough information about the prior locations of the planets. This ultimately led to a startling change in the way people perceived the surrounding world. They began to realize that there was an underlying order and elegance within the universe, where before all they had seen was chaos and peril.

Although mathematics did not become inextricably linked with science until the time of Copernicus and Kepler, this development had its seeds in the conversations and writings of the ancient Greeks. The Greek spirit of vigorous debate and inquiry was a model for all later investigators: it encouraged divergent viewpoints supported by logical arguments. Even though the Greeks had nothing akin to the scientific method, their clear-headed approach and openness to the investigative possibilities offered by mathematics separated them from all others. Whereas most ancient civilizations were content to use mathematics for everyday commerce, the Greeks created and used trigonometry to determine the distances to nearby celestial objects such as the moon and the sun. Their calculations showed that the moon and sun were much farther away than even the boldest of thinkers had imagined. This realization in turn prompted them to begin to wonder about the structure and the organization of the universe itself.

One disadvantage of living in ancient times was that there was little organized scientific data. Of course the absence of a developed science made it possible for some thinkers to offer outlandish thoughts about the world and, indeed, the cosmos as a whole. So one could assert that the world was balanced on the back of a giant tortoise or that the sun would be swallowed up by a huge bird every evening and regurgitated the following morning—and still not be precluded from attaining a position of status in society, such as the imperial astronomer. After all, there was no systematic collection of scientific observations and records that could be used to refute such ideas in a logical and straightforward manner.

But the Greeks had begun to change the anarchy that prevailed in the world of primitive science—largely by their use of mathematics. Indeed, it was their use of trigonometry that compelled them to conclude that earlier views of the heavens as a black shroud hanging over the earth could not be sustained. After all, the sun was found to be some 90,000,000 miles away. The fact that mathematics could dispassionately provide so much accurate information led the Greeks to wonder if mathematics could also be the key to understanding the dynamic motions of the celestial bodies. More than any other ancient civilization, the Greeks were very interested in trying to explain the movements of the stars and the planets in the evening sky in purely physical terms; they were not merely content to attribute the phases of the moon or the changes of the season to the moods of the gods. Some of the more intellectually curious seemed to believe that there had to be more fundamental, more primordial explanations for the varied phenomena that manifested in the world around them.

One of the first brilliant philosophers seeking physical explanations for the observed motions of the celestial bodies was one of Plato's students, Eudoxus, who offered a brilliantly innovative—though flawed—model of the cosmos that envisioned the sun as a solitary orb around which the planets and other stars traveled. Fortunately, Eudoxus was not executed for daring to suggest that the earth might not be the center of the universe. Most of his fellow Greeks were not concerned about whether the earth circled around the sun because they were more worried about finding enough food or being killed in the on-going wars between the city-states. But Eudoxus's theory did attract some support among other Greeks, due in no small part to his reputation as the preeminent mathematician of that era—even though his theory eventually fell into obscurity.

Now it would be nice to say that Eudoxus's theory was quickly plucked from the dustpile of history and restored to its rightful place in the pantheon of scientific ideas, but the plucking took nearly two millennia. Moreover, Eudoxus's theory, which had been offered to explain the observed movements of the sun, moon, and planets, was not completely accurate and needed to be overhauled. As a consequence, the effort to use geometric principles to model the universe based upon the idea of the planets moving in circular orbits around a common point changed over time. The earth instead of the sun was placed at the center of the universe.

The two most influential astronomers of the ancient era—Hipparchus and Ptolemy—were determined to construct a model of the cosmos that was grounded on mathematical principles. Like earlier astronomers, they needed to explain why an observer on earth looking up in the sky would see some planets moving forward and others appearing to travel backward. To account for this visual anarchy, they proposed an earth-centered model of the universe laden with epicycles and deferents. In this model, the sun moved around the earth on a circular path called the deferent; the various other planets moved around the sun as its spun around the earth, following smaller orbital paths called epicycles. Needless to say, this model was extremely cumbersome. In the original model offered by Hipparchus, each orbiting body had several epicycles, "each of whose centers moved on the next one and a final one on a deferent."[3] A complicating factor was that Hipparchus and Ptolemy had to construct these orbital schemes in order to fit the observed motions of each orbiting body. The model had to be assembled piece by piece with each set of epicycles specially designed to account for the movements of a particular heavenly body. This was not the highwater mark of science from the standpoint of creating a theory from which verifiable predictions could be derived. Indeed, its orientation was essentially inductive in which the observations made by astronomers were assembled and a mathematical model constructed to account for the observations. However, it was a very impressive effort to describe the universe mathematically even though it was very awkward. And while it did not precisely explain the dynamic motions of the heavenly bodies, it did underscore the inextricably close relationship between mathematics and science—even though this relationship was not readily apparent to everyone at the time.

This geocentric model, despite its shortcomings, was very durable and remained a dominant paradigm of human thought for nearly fifteen centuries. But its longevity was due as much to the collapse of the western civilization (as the Roman Empire became overrun by barbarian tribes from the East) as it was the worth of the theory itself. As barbarians are not known for their interest in astronomy, their presence did not help to encourage further inquiries by scholars into the merits of the geocentric model. More important, however, was the fact that the so-called Ptolemaic model was bolstered by the calculations of some of the greatest mathematicians of antiquity; this intellectual pedigree discouraged others from actively challenging the model. It is probably also true that the collapse of the Roman Empire caused society in general and individuals in particular to think less about grand questions such as the structure of the cosmos and instead focus on more immediate concerns such as trying to survive in a chaotic, dangerous world. After all, you are not likely to ponder the intricacies of planetary orbits when you are being chased by a sword-wielding barbarian.

The rise of the Christian Church over the ensuing centuries also discouraged scientific inquiry because issues regarding the position of the earth in the universe were of pivotal importance to the church leaders: Any suggestions in favor of a heliocentric (sun-centered) theory of the universe seemed to challenge the correctness of biblical doctrine and, hence, the primacy of the church in world affairs. Needless to say, those who dared to question the primacy of the geocentric theory were often incarcerated and, on more than a few occasions, tied to a stake in a village square and burned. It is one of the curious paradoxes of history, however, that the church's reliance on a literal interpretation of the Bible for its orthodoxy necessitated that it adopt an account of the creation and structure of the universe that was originally created by the Babylonians. To prop up the geocentric theory so dear to its religious doctrines, the church disregarded many aspects of the Greek cosmology, including the idea that the world is round. Instead the church leaders accepted as truth many ideas that even most young children now know to be wrong-headed, such as the idea that the earth is flat. There was no interest in actually verifying whether these statements were true or not; they were simply accepted at face value. Scientific inquiry as we know it—in which an investigator would actually determine whether there were facts to back up an assumption—simply did not exist. The average person merely accepted whatever was told to him by the local ecclesiastical authorities. Most Europeans had little knowledge of the Greek classics and this ignorance continued throughout much of the medieval era. It was only after these works were reintroduced many centuries following the fall of Rome in 476 C.E. by Arab scholars that many Europeans began to consider seriously such things as the position of the earth in the universe.

But we should not assume that a growing familiarity with the Greek classics in medieval Europe brought about widespread acceptance of the heliocentric theory. It was still quite uncommon for many persons to concern themselves with anything that did not affect their daily survival. And it is one of the curious features of history that the heliocentric theory, as we have noted previously, was revived single-handedly by an obscure Polish astronomer and clergyman, Nicolaus Copernicus. What makes Copernicus's embrace of the sun-centered theory so surprising is that he was one of very few persons in the early sixteenth century who believed that the Ptolemaic model of the cosmos was fatally flawed. Moreover, he was willing to devote much of his life to developing a more elegant heliocentric model. Copernicus was akin to a rebel in the wilderness because it is far more typical for science to evolve in fits and starts, with one idea leading to another, and occasionally an entirely new paradigm taking another's place or even pushing aside the existing orthodoxy. But in the development from geocentricity to heliocentricity, one could not gradually evolve from an earth-centered to a sun-centered cosmology—one basically had to make a leap of faith and decide in favor of one or the other. In spite of or perhaps because of his training at the University of Bologna where he studied the Greek classics and learned to make astronomical observations, Copernicus began to question the prevailing assumption of the era that all of the planets and stars revolve around the earth.

After studying both law and medicine, Copernicus was appointed canon of the Frauenberg cathedral in East Prussia, where he served as an administrator. This was a particularly fortunate turn of events for Copernicus because it provided him with ample leisure time to carry out his astronomical investigations. In his spare time, he made many observations of the evening sky that, over the course of several decades, led him to the conclusions offered in his book *On the Revolutions of the Heavenly Spheres*, which was published while he lay on his deathbed in 1543.

The essential genius of Copernicus lay in his belief that the Ptolemaic model of the universe had become so cumbersome and complex that it could not possibly be correct. In this way, Copernicus seemed to share a belief later offered by Einstein that the universe should be governed by simple principles. Indeed, the observations he made of the evening sky gradually convinced him that he could clean up many of the needless complexities of the Ptolemaic model by moving to the comparatively simpler heliocentric model. This was a conclusion for which we should all be grateful to Copernicus; it represented one of the first times in the history of modern science in which mathematics was used as a tool to reveal both fundamental truths about nature and to devise a more elegant yet easier to understand theory of the world.

Why was the Ptolemaic theory so complicated? Aside from the difficulty many people have in pronouncing the word "Ptolemaic," the classic geocentric cosmology proposed that the earth was at the center of the solar system. In this system, as we have pointed out, the sun circled around the earth on an imaginary orbit defined by a circle called the deferent. The other planets in turn circled around the sun in their respective orbits called the epicycles as the sun traveled along the deferent. Although this theory was widely adopted, it did have the drawback of being completely wrong (but its early proponents did at least manage to avoid making the erroneous assumption that the earth is flat). It would fall to later scholars and theologians in the medieval era to compound further the inaccuracies of the Ptolemaic theory and drop the idea of a spherical earth in favor of one that was flat. This was surely not one of the finer examples of the continued evolution of scientific thought over the course of the centuries.

No doubt you are wondering whether the debate over epicycles, deferents, and Ptolemaic cosmology ever lead to a European war or even a military skirmish. Sadly, debates over the finer points of cosmology and astronomy rarely provoke the reaction that prompt entire populations to mobilize for war. But for those persons who do care passionately about whether the earth moves around the sun or the sun moves around the earth, the debate over the correctness of geocentric and heliocentric theories was of paramount concern. Certainly Copernicus believed very strongly that it was absurd to assume that a comparatively small planetary orb such as the earth could be the fulcrum around which the entire canopy of creation turned. But Copernicus was also well aware of the penalties that were meted out to those who dared to question the Christian Church's position on earth-centered cosmology, which they favored. The Italian mystic Giordano Bruno, for one, had been burned to death for daring to suggest that the universe consists of an infinity of worlds. Copernicus obviously learned something from Bruno's ill-advised comments and purposely delayed the publication of his book summarizing his researches as long as possible to avoid a similar fate. But he was not one to keep entirely quiet about his views and so he authored a little tract called the *Commentariolus* that was secretly passed among a few trusted friends and colleagues. Although his friends urged him to prepare a full-length work and offer a more detailed view of his heliocentric theory, Copernicus remained reluctant to undertake the task lest he be literally tossed on the bonfire of history. Certainly it was easier for Copernicus's friends to be enthusiastic since they would be spectators instead of the main entertainment at any public burning. But Copernicus's tract did make it possible for his views to be made known to a number of prominent theologians and scholars and they, in turn, provided Copernicus with some constructive criticisms that helped him to refine further his ideas.

By proposing that each planet moves in a separate orbit around a stationary sun, Copernicus was able to dispense with many of the mathematical complexities of the geocentric model and still explain many of the observed motions of the planets in the evening sky. This was not an obvious solution because all astronomers were faced with the task of having to explain the seemingly erratic motions of the planets in the evening sky: some of the planets appeared to move in one direction and then move back in a reverse direction across the evening sky. By dropping the assumption that all of the other planets were moving around a sun that was in turn circling around the earth, Copernicus was able to get rid of many of the epicycles that plagued the Ptolemaic model. This further justified in his own mind tossing aside the geocentric cosmology.

It would be a fitting ending to say that the Copernican theory solved all the problems dogging the astronomers of the day and provided a perfect blueprint to explain the movements of the planets. But the Copernican model was not much more accurate than the model offered by Ptolemy because Copernicus had assumed that the planets move around the sun in perfect circles. We now know that this is not the case as the planetary orbits are elliptical, appearing as elongated ovals. Copernicus was so impressed with the mathematical symmetry of his heliocentric model of circular orbits that he could not imagine an alternative model. But he did recognize that using perfect circles did not completely explain the motions of the planets. He found himself having to patch up his model by adding a few epicycles here and there—though not so many as had been tacked on by Ptolemy. Sadly for Copernicus, however, this new and improved model was not much more accurate than his original heliocentric model. Copernicus did not allow these discrepancies to cause him to question the basic premises of his model; they instead served to motivate him to continue tinkering with his heliocentric model for the rest of his life in the ultimately futile hope that he would be able to iron out its inconsistencies.

If we measure success by the amount of attention a matter receives, then we could characterize Copernicus's *On the Revolution of the Heavenly Spheres* as being a public relations magnet; it attracted the attention of both university scholars and church leaders alike. Because Copernicus had skillfully managed to die on the day his book was published, he could not be put to death by the church. But in a Europe seething with political turmoil and religious strife, the church decided that the allegedly inflammatory conclusions of Copernicus's work were not worthy of consideration by the public. It thus banned the publication of Copernicus's book as well as any books making reference to it. Indeed it was not until the early nineteenth century that the church reconsidered its position and decided that Copernican heliocentricity could be studied.

Although Copernicus surely ranks as a heroic figure in the history of science, his model was at best incomplete. It fell to the German mathematician Johannes Kepler to use his own knowledge of mathematics to refine Copernicus's model so that it would accurately describe the movements of the planets. Kepler himself had studied for the ministry as a young man but his lack of patience with the unflinching orthodoxy of the church eventually caused him to drop religion in favor of astronomy. While studying astronomy as the University of Tubingen, however, Kepler first became acquainted with the work of Copernicus. Although he considered himself to be a devout man, Kepler, like Copernicus, did not feel that a belief in God required one to follow all of the dogmatic rules of the church.

Because Kepler himself was something of a mystic, he, too, was captivated by the circular orbits predicted by Copernicus. Indeed, his conviction that the heavens were the handiwork of a Divine Creator caused him to spend many years trying to devise an accurate model of the solar system with the circular orbits proposed by Copernicus. But he ultimately concluded that the circles, despite their geometric perfection, were simply not the correct shapes for planetary orbits. Now Kepler could have taken a more radical approach and used triangles or squares but he took his time, convinced that only a few subtle changes were necessary. In 1609, Kepler published a book that examined the orbit of the planet Mars and offered the first two of his three laws of planetary motion. The first law, which represented a break with the Copernican circular orbits, stated that each of the planets move in elliptical orbits around the sun. Kepler found, much to his delight, that the use of elliptical orbits enabled him to jettison the epicycles and deferents that had ruined the earlier cosmological models and serve up a model that could accurately describe the observed motions of the planets.

Kepler's second law came from a thoughtful study of the ellipse itself and the velocity of the planets moving around the sun. Copernicus, like earlier astronomers, had assumed that the planets move in constant velocities around the sun. If you are a proponent of circular orbits, then this would appear to be a sensible conclusion. But if you switch to an elliptical orbit, then the movements of the planets become more problematic. Kepler discovered his so-called law of equal areas, which states that a planet moving in a given orbit sweeps out over equal areas in its orbital path in equal times. This means that the triangular space formed by the extension to the sun of the endpoints of the orbital segment covered by the planet in a given time at its farthest position from the sun is the same as that covered by the same planet at its nearest proximity to the sun. Because the planet would increase in velocity as it moved toward the sun and decrease in velocity as it moved away from the sun, the dimensions of the base and altitude of these imaginary triangles would continuously vary, but they would always be equivalent for equal periods of time. This law convinced Kepler that mathematical principles play a pivotal role in the structure of the universe and that the simplicity of these concepts was evidence of the presence of a Divine Creator.

The alert person will notice the reference that was made to Kepler's three laws of planetary motion. Many persons would be very pleased to have picked up just two of the three laws, but no discussion about mathematics and astronomy would be complete without an account of Kepler's third law. This law proposed that there is an inverse relationship between the distance of a planet from the sun and the period of time it takes to revolve around the sun. This law may be expressed as follows:

$$kD^3 = T^2$$

where k is a constant, D is the distance of the planet from the sun, and T is the time it takes for the planet to revolve around the sun. This law seems to be a very simple algebraic relationship but it is difficult to overstate its importance to both mathematics and physics. Why? Kepler's third law offered a mathematical expression that described the direct relation between the distance of a planet from the sun and the amount of time it took to complete one revolution around the sun. As such, it hinted at the role played by gravitation in keeping the planets in the solar system from spinning off into space. But Kepler knew nothing of the law of universal gravitation that would later be proposed by Isaac Newton, although he probably suspected that there was some sort of unseen force that was holding the planetary system together. In any event, Kepler's third law also underscored how mathematics could be used to describe purely physical phenomena and, more importantly, how it could be used to predict planetary motions. This discovery's importance was not lost on Kepler, who believed it provided proof of a fundamental harmony in the universe.

Harmony? Music harmony? Well, Kepler was something of an enigma. His search for general principles to govern the universe had also led him to allow his imagination to wander in incredible flights of fancy. Indeed, Kepler went so far as to search for a relationship between the angular velocities of the planets and the notes on a musical scale. But even though he stumbled upon a few interesting coincidences, these efforts were largely unsuccessful. But he was very happy with his third law, which first appeared in his 1619 book, *The Harmony of the World*. Indeed, he viewed it as the culmination of a career, which—despite occasional detours into mysticism and astrology—guaranteed Kepler a place of honor in the pantheon of great thinkers.

Kepler's laws of planetary motion are arguably a cornerstone of modern science and provide a mathematical underpinning for Copernicus's daring vision of a heliocentric solar system. But his search for these laws consumed most of his energies throughout his career. Kepler worked as a professor of mathematics at the University of Graz in Austria, but often gave astrological readings in order to make ends meet. It was only after he obtained a position working as an assistant to the famed Danish naked-eye astronomer Tycho Brahe that Kepler was given access to the astronomical tables that would provide the empirical data he needed. These data were critical to his formulation of his three laws of planetary motion. Although Kepler succeeded Brahe as the Imperial Mathematician to Emperor Rudolph of Bohemia in 1601, the kingdom suffered from the financial troubles that most storybook kingdoms are able to sidestep. Kepler himself took a position as a mathematician in Linz in 1612 but his salary was small and he was barely able to stay ahead of his creditors. His wife and three of his children died over the next few years, and his own health continued to deteriorate until he died in 1630.

It is difficult to appreciate the adversity that Copernicus and Kepler had to overcome in order to carry out their scientific work. Their conclusions stood in complete opposition to the prevailing orthodoxy of the church and most respected scholars. Not only did they risk professional humiliation but they also had to be mindful of the dangers of expressing their views too stridently lest they be seen as heretics and subject to one of several rather unpleasant forms of torture or even death. Certainly the religious leaders of the time were not reluctant to toss a few freethinkers onto the bonfire in order to underscore the correctness of their views. With the spectacle of punishments for daring to question the prevailing wisdom being extreme, it required great fortitude to follow one's intellectual convictions and support something as revolutionary as heliocentricity.

But the opponents of heliocentricity did ask some insightful questions that neither Copernicus nor Kepler could answer: What was the nature of the force that caused the earth to begin moving through space in the first place? (Here the answers invariably got bogged down, as they do today, in responses couched to varying degrees in religious and scientific terms.) Why were the objects on the surface of the earth not thrown off by the spinning earth in the same way that a stone tied to the end of a twirling string is continually flung outward? (Copernicus and Kepler had no knowledge of the gravitational force that binds people and objects alike to the surface of the planet. They would have greatly appreciated an appearance by Newton to solve this question.) A more practical astronomical question involved the apparent lack of movement of the stars that should be visible if the earth were revolving around the sun. Here, the basic conclusion that the stars did not move was made only because the shifts in the positions of the stars were very difficult to discern with the naked eye. But perhaps the most pedestrian argument that any person could make against the idea of a moving earth was the absence of any innate sense of motion. In other words, we are not constantly falling down or trying to steady ourselves as might be the case if we were standing on a merry-go-round or a bus. This example would have carried more weight in Copernicus's day if they actually had merry-go-rounds and buses, but it illustrates a point that troubled both Copernicus and Kepler.

Even though Copernicus and Kepler had many doubters and were not able to respond satisfactorily to every query directed at them, they were surprisingly confident that their views would be vindicated by posterity. This conviction was prompted by their steadfast belief that the heliocentric theory represented a much simpler, more elegant model that was more worthy of their Creator than the unwieldy but more mathematically developed geocentric theory. This view was all the more surprising when we consider that the mathematical underpinnings of the geocentric theory had been developed for more than a millennium, whereas the heliocentric theory was the equivalent of a poor stepchild with very little mathematical clothing until the time of Kepler.

Kepler's work in astronomy forever joined him at the proverbial hip with Copernicus and set the stage for the great achievements that would be made by Galileo Galilei and Isaac Newton in mechanics, physics, astronomy, and mathematics. But the achievements by Copernicus and Kepler may have required even more courage and intellectual daring than those of their more famous successors because they lived in a time in which science, astrology, mysticism, and mathematics were all knotted together and the process of scientific investigation was generally unknown. Fortunately for science, the simplicity of the sun-centered theory—which appealed to both scientists and even some religious scholars—helped to show that a universe of orbiting worlds where even the earth itself whirled around a sun was not only possible but also probable. This vision was coupled with Galileo's observations of the motions of the other orbiting planets and moons in the solar system. The constantly shifting positions of the planets also underscored the falsity of astrology, which itself had ascribed great significance to the apparently fixed positions of the stars in order to divine the future. The Copernican cosmology thus caused many persons to wonder whether the stars themselves move through space—like their planetary counterparts.

thirteen

One good thing about living several centuries ago was that there were so many wonderful theories and ideas yet to be discovered. Indeed, there may still be many such undiscovered treasures waiting to be identified. But very few of us are lucky enough to be blessed with the intellectual powers needed to ask the right questions and make the appropriate observations at the most opportune time in history. Even those of us who are brilliant enough to be first-rate scientists and philosophers are often too preoccupied with such matters as trying to make a living rather than devoting our lives to creating a breakthrough theory of the universe.

Yet the human race still manages to produce persons of genius in every generation or so who somehow change the ways in which we view the universe and our place within it. And a few of these geniuses possess such astounding intellects that they are able to make important contributions in two or more separate fields of knowledge. When we hear the names of Descartes and Galileo, for example, we know that we are talking about two such men. After all, modern science and mathematics—indeed our modern society as a whole—would not have been possible without the mathematical and scientific breakthroughs ushered in by these two giants of the Renaissance. So now that we have a better understanding of mathematics, it is important that we return to these brilliant thinkers to consider in detail some of their contributions to our knowledge of the physical world.

René Descartes (1596–1650) was born into a well-to-do family in La Haye, France, and received his formal education at the Jesuit College at La Fleche. He then studied law at the University of Poitiers. However, Descartes was not drawn to the study of law, which he viewed as ossified and reactionary. Instead he became consumed with cosmology and studied the thousand-year world view offered by Ptolemy that had shaped European thought to that point in time. From the very beginning, Descartes had a skeptical world view and seemed to take a delight in challenging the correctness of almost any widely held assumption. But before Descartes concerned himself too much with the weightier issues inherent in developing a new philosophy and, incidentally, creating a couple of new branches of mathematics along the way, he did decide to "find himself" in the pleasures of the flesh in Paris. And while it would certainly be more interesting to read about Descartes's lurid adventures in Paris, the true mathematical student will instead insist on moving ahead to discuss Descartes's achievements in mathematics. Suffice to say, Descartes enjoyed himself for a few years, then swung to the opposite extreme and spent some time in contemplative study before joining the army and participating in several military campaigns. It was only after Descartes began the fourth decade of his life that he decided to devote himself to the task of constructing an entirely new philosophy based upon certain immutable axioms much akin to the axioms that underpin Euclid's geometry.

Let us first consider René Descartes's significant contributions to philosophy before moving on to his mathematical endeavors. After all, his personal philosophy was the prism through which he viewed the world and, presumably, influenced the manner in which he approached mathematical problems. Descartes believed that truth could be found only through the use of reason. Knowledge was therefore the culmination of a rigorous intellectual reasoning process. It was the process of thinking as embodied in his famous quote ("I think, therefore I am") that led him to seek knowledge through the use of basic propositions and ultimately led him to offer detailed arguments regarding both the existence of God and the physical universe itself. Descartes's notion of self-actualization (e.g., "I think, therefore I am") has invited scholarly debate over the concepts of thought and awareness, which continue to be hotly debated in the universities to this day.

Descartes also made important contributions to mathematics, particularly the field of analytical geometry (which is the study of geometrical relations by means of the techniques of analysis).[1] However, it was his impatience with the limitations of Euclidean geometry, as we have discussed elsewhere, that prompted him to integrate algebra and geometry to create a coordinate system that consists today of an x-axis (horizontal) and a y-axis (vertical). (We should point that the French mathematician Pierre de Fermat arrived independently at many of the same conclusions as Descartes.) As discussed earlier, each axis is in turn numbered with whole numbers, with the intersection of the two axes being the value 0. With 0 as the point of origin, the units would increase uniformly on the x-axis as one moved to the right of the 0 (1, 2, 3, 4, and so on) and decrease uniformly on that same axis as one moved to the left of 0 (1–, –2, –3, –4, and so on). The same numbering system was used on the y-axis so that the units would increase uniformly as one moved above the 0 (1, 2, 3, 4, and so on) and decrease uniformly as one moved below the 0 (–1, –2, –3, –4). One could use this coordinate system to plot the location of any point on the plane formed by the intersection of the horizontal and vertical axes. So if one wanted to specify the location of a particular point on this plane, he or she would simply move along the requisite number of units on the x-axis and then repeat the process along the y-axis. A point located at position "A," for example, would be described by moving 3 units to the right of the point of origin on the x-axis and then moving upward 2 units on the y-axis. Its location would then be specified as (3, 2), which is a shorthand way of specifying its x and y positions, respectively.

Coordinate systems have been widely applied in everyday life. Indeed, many of us who have tried to find a remote town on a map have made use of a coordinate system because the horizontal and vertical edges of most roadmaps are marked in letters and numbers that correlate to an index of town names. So if we are planning to visit an art museum in Scranton, Pennsylvania, for example, we may have to look at an index on a map of Pennsylvania that may say, for example, that Scranton is located in the square on the map that is "C" spaces to the right of the lower-left corner and "7" spaces above the lower-left corner. The fact that a map may use letters instead of numbers on one of the axes does not really make any difference, because both letters and numbers can serve the same function as long as the units increase in a uniform and predictable manner (e.g., A, B, C, etc.)

The coordinate system is not limited to purely mathematical applications even though we do not often see negative coordinates used on maps. This is not to say that you could not include negative coordinates and have the intersection of the x- and y-axes at the center of the map, but most people would find such a feature to be overly complicated. No doubt a person reading a map could figure out the location of a town located at –B, –4, for example, by moving to the lower-left quadrant of the map. But it is conceptually simpler to place the point of origin (0) at the lower-left corner of the map and then mark off the horizontal and vertical axes accordingly.

This coordinate system could also be used to represent three-dimensional objects, such as the terrain described by a topological map. One needs only to draw a third axis, the so-called z-axis, perpendicular to the existing x- and y-axes. In this way, one could describe the position of a point in space by marking its distance from the point of origin in terms of three coordinates—its three positions on the x-, y-, and z-axes. The beauty of this system is that there is only one unique point located at any given intersection of the three axes. As a result, I can say that a point can be found merely by specifying the values of each of its three axes, which, in this case, are 2, 4, 6, and represent the x-, y-, and z-values, respectively.

The idea of using a coordinate system has influenced many areas of our modern society because coordinate systems are used in everything from cartography to astronomy. But this rudimentary coordinate system, useful as it may be, was only the starting point for the work of Descartes because his venture into geometry was motivated by his desire to discover a better method for proving theorems about curves—whether they were the orbits of the planets, the paths followed by projectiles such as cannonballs, or even the trajectories of moving objects. In any event, Descartes, having developed the coordinate system that would bear his name (Cartesian coordinate system), then began to examine the properties of curved lines in a coordinate system. He first plotted points of varying x- and y-values on his coordinate system. He found that as he reduced the amount by which he moved along the horizontal and vertical axes, he could place more and more points and more closely approximate a curve. In short, he was trying to understand curves by working with successively smaller straight line segments. As a result, he could move down the horizontal axis along the points x_1, x_2, x_3, and then plot corresponding points y_1, y_2, y_3 on the vertical axis, and thereby create a curve by simply "connecting the dots." Now Descartes was a busy fellow who did not have time to sit hunched over his desk and plot out hundreds upon hundreds of points on a sheet of paper to create intricate curved lines on his coordinate system. So he wanted to create a more generalized expression that would provide a short-cut method and thereby avoid the drudgery of trying to draw the infinite number of points that would be contained in any line—curved or straight.

　　Descartes found the key to this problem by returning to algebraic notation and setting up various expressions with x and y values whereby the definition of one value would automatically establish the other value. The simplest algebraic expression involving both the x- and y-variables (in which we are selecting values for both the x- and y-axes and then plotting the point at their intersection) is the expression $x = y$. So if Descartes defined x as being 1, then y would also be equal to 1. If x were 2, y would also be 2, and so on. After setting three or four values for x, Descartes would then have three or four corresponding values for y and could then plot the points at the following locations (1, 1), (2, 2), (3, 3), (4, 4), which would be a straight line moving from the origin 0 at a 45 degree angle.

　　But any mathematical student knows that a straight line is not a curved line. Still, this algebraic expression would provide an equivalency relationship that would hold up regardless of the value that was chosen for x. Descartes could set the value of x equal to 9, 23, 748, or even 2,456,987,233 and the equation would automatically define the corresponding y-value and thus the location of that particular point. So even though an infinity of numbers could be plugged into this equation, Descartes could plot this expression by finding only a few values for x, determining the corresponding values for y, and then plotting the locations of these points. He could then draw a line through each of the points and essentially connect the infinity of points contained within that line segment.

　　But this was only the beginning; it was mere child's play to draw a straight line when one could draw many interesting curved lines. One merely had to select the appropriate algebraic equation and then charge ahead. One of the curved lines that caught Descartes's fancy was that of the parabola, which can be expressed with a variety of equations so long as one variable (such as x) is expressed as a first-degree variable (e.g., x, $2x$, $x/2$, etc.) and the other is expressed as a second-degree variable (x^2, $3x^2$, $1/5x^2$). So Descartes could select the equation $x = y^2$ and, by substituting the following y values (–4, –3, –2, –1, 0, 1, 2, 3, 4) would obtain the following values for x (16, 9, 4, 1, 0, 1, 4, 9, 16). These would result in a set of values that would look like an upside-down horseshoe.

Not only does this particular expression describe a parabolic curve that moves upward without limit from both the negative and positive sides of the point of origin but it also neatly represents with just a couple of symbols a mathematical relationship that could include literally countless individually plotted points. So it is not an exaggeration to say that perhaps the greatest achievement of symbolic notation is that it enables us to express in a very abbreviated form with ever-more exacting placements of plotted points certain mathematical relationships that one could spend a thousand years trying to describe graphically.

The fact that curves of all kinds could be represented algebraically and thereby studied in a conceptual manner was one of Descartes's (and Fermat's) greatest contributions to mathematics. One did not have to describe a geometric figure graphically and then try to break it down using the laborious methods of the Greek geometers. Instead one needed only to master the subtleties of algebraic equations so that one could thereby describe any curve mathematically and uncover the properties of that curve through conceptual manipulations of the mathematical equations themselves. In a way, it was as liberating as learning that one did not have to count from one to one million one by one in order to get to one million; if one knew how the counting process worked, then one did not need to count every single number to arrive at one million.

But Descartes, being the curious soul that he was, did not content himself with describing curves alone. Indeed, it was only a matter of time before he decided to expand his mathematics to the study of surfaces. Of course this placed entirely new demands on his algebraic equations. After all, a curve could be represented in a single two-dimensional plane whereas a surface could only be represented only in a three-dimensional space. But Descartes was not one to shrink from a challenge and he was soon inspired to add a third axis to his x-y coordinate system, so that it consisted of three mutually perpendicular lines. This may seem difficult to visualize but we could picture these lines as the intersection of two walls in a house with a floor so that the corner on the floor would represent the point of origin (0). The two lines formed by the walls and the floors would represent the x- and y-axes, and the third line formed by the intersection of the two walls themselves would represent the third so-called z-axis.

It is all very well to talk about three-dimensional spaces, but how does one go about using algebra to describe the three-dimensional surfaces in these spaces? The answer is very simple. After all, the two-dimensional coordinate system has two sets of variables that describe coordinate positions on the x and y axes. Not surprisingly, the three-dimensional coordinate system has three sets of variables that in turn denote coordinate positions on the x-, y-, and z-axes. So the position of any point in space can be described only if we know its three positional coordinates that correspond with their respective x-, y-, and z-axes. So if we want to locate, for example, point A in three-dimensional space, we have to determine its three spatial coordinates and then plot the one point in space where they all intersect. Using our algebraic shorthand, we would say that point A is located at (2, 3, 5), the numbers refer to its coordinates on the x-, y-, and z-axes, respectively.

In trying to offer an algebraic description of a surface such as a sphere, Descartes realized that he would have to incorporate all three variables into his equation. He knew that he was describing, in a shorthand form, the plotting of all of the points in a three-dimensional space that were the same uniform distance—say, 6 units—from the center point. By reworking the Pythagorean theorem, Descartes derived an equation that described the sphere (using 6 units) as follows: $x^2 + y^2 + z^2 = 36$. This equation describes the sphere only when the equality holds and the three coordinates of a point located on the surface of the sphere are substituted in place of the three variables. The important message is that

Descartes was instrumental in the development of methods to describe mathematically all types of surfaces; this in turn made possible a deeper understanding of many of the properties of these surfaces. These methods were also further developed to deal with higher-dimensional spaces and, ultimately, would play a role in Albert Einstein's special theory of relativity, which posited a four-dimensional space-time continuum consisting of three spatial dimensions and a fourth temporal dimension.

Although Descartes helped to bring algebra into the modern era by fusing it with geometry, it was his very popular philosophical writings that won him widespread popularity throughout Europe. He spent the most productive years of his life in Amsterdam, free to think about the fundamental nature of truth, the universe, and the nature of god. But his seclusion was interrupted when he was summoned by Queen Christina of Sweden to serve as her personal tutor. Although Descartes was reluctant to leave Holland, he could not resist the trappings of the royal court, and so he accepted the offer to come to Stockholm, expecting to enjoy a leisurely lifestyle at the royal court. But Queen Christina, who boasted an impressive intellect as well as great physical abilities, had other ideas, and, apparently, very little need for sleep. In fact, Descartes was soon dismayed to discover that he was expected to meet the queen every morning at 5:00 sharp in an unheated library and guide her through a grueling set of lessons for much of the day. For the queen, learning was a pleasure; moreover, she eagerly absorbed the lessons of her professor. But the tutoring exacted a dreadful toll on Descartes's frail constitution; Descartes lasted scarcely two years in Stockholm before succumbing to a pulmonary disease in 1650.

ENTER GALILEO

Although Descartes made great contributions to mathematics and philosophy, it was the Italian astronomer and physicist Galileo

Galilei who played the most indispensable role in the evolution of science, elevating it from the murky, mystical morass in which it had languished for thousands of years. Galileo was the first to decide that the goal of science should be to use mathematics to quantify the phenomena of nature instead of trying to pursue ultimately unsolvable philosophical questions as to why events occur in nature. Indeed, the Greeks, including Aristotle, were notorious for offering reasons for the existence of certain phenomena that offered few insights into their actual workings. They believed, for example, that the planets move in circular orbits because circles are perfect geometric figures and thus the only form appropriate for describing the paths of the heavenly bodies through the void. Little thought was given to the physical forces that might cause the planets to travel in circular orbits. Things did not change much in the medieval era except that the reasons for the movements of the planets or the changes in the seasons or the variations in the weather were now attributed to God. Even though this might not seem like an objectionable viewpoint, it did not really help to advance the progress of scientific thought. To simply explain all things as being part of God's plan discouraged people from trying to understand and explain the dynamic processes in the world around them. Thus many questions about the physical world were not actively pursued because there was no interest in understanding the nature of the phenomena themselves at that time in history. Indeed, it was thought to be more important to explain all phenomena in terms of their particular places in the prevailing religious orthodoxy.

Perhaps Galileo's greatest contribution to both mathematics and science was the ability to describe observed phenomena in a quantitative (purely mathematical) manner and not bother with the basic philosophical questions that had ensnared most scientists and mathematicians before him. In his studies of motion, for example, Galileo would drop weights from great heights and try to calculate the distance these weights fell with each passing moment. Aristotle and his fellow Greeks would have become involved in long-winded debates as to why the objects were falling to the ground and seeming to move faster and faster with each passing second. Galileo did not concern himself with the mystery behind the falling weights. Instead he sought to use mathematics to describe the changes he observed in the speeds of the falling objects. He saw that there were two variables at work in these experiments: the passage of time itself and the velocities of the falling objects. For Galileo, the challenge was to create an algebraic equation that would express the relationship between the distance traveled by the falling object and the time that elapsed during the fall. Galileo was guided in this endeavor by the results of his own experiments that had shown there is a direct relationship between the passage of time and the distance traveled by the falling object. Furthermore, his experiments had demonstrated that the velocity of any falling object increases with time. In a way, the acceleration of these objects complicated Galileo's attempt to formulate an algebraic expression to describe the motions of the falling objects. His observations and his skill with mathematical calculations enabled him to derive a formula that is generally expressed as $d = 16t^2$ where d is the distance traveled and t is the time. In short, this formula states that the number of feet that an object falls is equal to the sixteen times the number of seconds squared. In the case of Galileo's dropping weights from high places such as the Tower of Pisa, this equation tells us the distance that a falling object will travel at any given point in time. After 1 second, the object will have fallen 16 feet; after 2 seconds, 64 feet, after 3 seconds 144 feet; after 4 seconds, 256 feet, and so on. Of course this relationship could not be observed for longer periods of time because Galileo did not have endlessly tall towers available from which to throw things. No matter how high he climbed, he was also limited by certain experimental and observational factors because the objects would eventually hit the ground. But Galileo's equation neatly and cleanly expresses the idea of acceleration and accounts for the effects of what Newton would later identify as the universal gravitational force. Galileo's equation can also be used to describe an infinite number of problems, depending on the values chosen for the variables. At the same time, it offers exact results once the values are chosen. It also offered the important benefit of saving Galileo and others from having to lug a lot of heavy objects up to the tops of any towers in order to collect additional observational data. Instead, he could simply plug in a number for t and then he would automatically obtain a quantity for d—the distance traveled by the falling object in the time t.

This equation showed how two different changing quantities could be related to each other and the way in which this dependency could be expressed. This equation expressed a functional relationship because it described how changes in the value of one variable are a function of changes in the value of the other variable. The word "function" is not an overly complicated concept. Indeed, it may be easier to think of functionality in terms of the dependence of one variable upon another. More specifically, the changes in one variable depend upon the changes occurring in another variable. As noted earlier, in describing a functional relationship, we may refer to an independent variable and a dependent variable. Not surprisingly, the independent variable is the one that undergoes the initial change. The dependent variable is the one that changes in response to the changes in the independent variable. To return to Galileo's example, the speed of an object tossed from the top of a tower would be a function of the elapsed time. The longer the object fell to earth, the faster that object would fall.

Functional relationships are not limited to mathematics because there are all sorts of functional relationships in the physical world. The distance that you can drive in your car depends on the amount of gas you put into the fuel tank. The more fuel you add, the farther you can travel in your car. In this case, the amount of fuel would be the independent variable and the distance traveled would be the dependent variable.

Having seen the huge amount of information that could be obtained by substituting different values for the variables in a simple equation such as $d = 16t^2$, Galileo became convinced that the overriding goal of scientists should be to find mathematical expressions that can succinctly describe the phenomena of the universe. Mathematics could then serve as a sort of universal language whereby seemingly disparate phenomena could be quantified and, hopefully, better understood: "[Such] formulas have proved to be the most valuable knowledge man has ever acquired about nature. . . . [The] amazing practical as well as the theoretical accomplishments of modern science have been achieved mainly through the quantitative, descriptive knowledge that has been amassed and manipulated rather than through metaphysical, theological, and even mechanical explanations of the causes of phenomena."[2] Indeed, the advent of mathematics has made it possible for the mystical, capricious natural world that was feared by the Greeks and the medieval Europeans alike to evolve gradually into a more logical system that could be probed and, to some extent, understood with the use of mathematics.

Galileo made numerous measurements of distances and elapsed times while carrying out his experiments with falling objects. Being an eminently practical person, Galileo did not try to answer the underlying philosophical questions relating to the direction of time or causality or any of a myriad of other profound questions that might have enthralled the Greeks. Rather, he was one of the first scientists to be as concerned with the accuracy of his measurements as the logical consistency of his theoretical arguments. Accordingly, he made great efforts to segregate those phenomena that could be measured from those that could not be measured. In this way, he separated himself from almost all of his predecessors who had attempted to categorize nature in terms of conceptual qualities such as essences and forms—things that could not be quantified. Certainly Galileo realized that he would not get very far trying to describe one-half of a form or the speed of an essence. So he focused his attention on such things like space, time, inertia, force, motion, velocity, and acceleration in order to put his mathematics to work. "In the selection of these particular properties and concepts, Galileo again showed genius, for the ones he chose are not immediately discernible as the most important nor are they readily measurable. Some, such as inertia, are not even obviously possessed by matter; their existence had to be inferred from observations. Others, such as momentum, had to be created. Yet these concepts did prove to be most significant in the rationalization and conquest of nature."[3] In this way, Galileo essentially framed the parameters of scientific inquiry and, in some sense, invented the modern scientific method. Here, he may have made a greater contribution to the evolution of science than even Isaac Newton or Albert Einstein. Indeed, Newton and Einstein could not have made their monumental discoveries in the morass of mystical, amorphous pseudoscience that dominated the thinking of seventeenth-century Europe—when Galileo began his investigations. The change in the nature of what we now call science wrought by Galileo was revolutionary. As with Copernicus in his reintroduction of the heliocentric theory, Galileo—in his adoption of a rigorous, quantitative, measurement-based investigative methodology—displayed almost unimaginable creativity. But we should not be surprised in some sense that Galileo (as opposed to Kepler or Descarates, for example), would be the one to establish the parameters within which scientific reasoning would be utilized. After all, Galileo was an eminently practical investigator who, while making important theoretical contributions to classical mechanics, was one who preferred to see and touch the things he was studying instead of merely thinking about them. So he was naturally less inclined to consider the questions as to why things occur as opposed to trying to describe their manifestations in a mathematical form. His equation regarding motion was undoubtedly rooted in theory but it was also influenced by his experiments with falling weights. So we have something of a chicken or egg type of question: Which came first, the theory or the observation? With Galileo it is not always clear, but it seems that his observations of the events around him played an important role in his formulation of his thoughts about motion. Or maybe it was the other way around. In point of fact, it is not really terribly important how Galileo came to derive his relationship between distance traveled and time elapsed for falling objects because he transformed science itself almost single-handedly.

The story of Galileo began with his birth in Pisa in 1564 into a family of some means. Even as a child, young Galileo showed great skill both in building toys and playing the organ. He also showed promise as a painter but seems to have been more interested in tinkering with mechanical devices than in placing brush to canvas. Although encouraged by his parents to pursue a career in medicine, Galileo was not particularly enamored with the prospect of a medical career even though he did study medicine at the University of Pisa. In 1584, while still engaged in his medical studies, Galileo discovered the law of the pendulum after noticing that each swing of the pendulum takes the same amount of time.

Over the ensuing few years, Galileo invented a device for measuring the specific gravity of objects and discovered his law of falling bodies, which states that all objects fall to earth at the same rate of acceleration, regardless of their masses. To buttress his arguments, he tossed objects having different weights from the top of the Tower of Pisa and carefully studied the outcomes of these trials. Although his law of freely falling bodies appeared to be vindicated by his experiments, Galileo's views evoked great anger among many persons who subscribed to the Aristotelian view that heavier objects should fall to the ground faster than lighter objects. For his part, Galileo, who had become a professor at the University of Pisa in 1589, refused to soften his conclusions in order to placate his foes and, as a result, was ultimately forced to resign his position.

After obtaining a position at the University of Padua in 1592, Galileo devoted himself to the study of astronomy. Having been convinced of the correctness of the Copernican theory, Galileo used his own skills as an experimentalist to build a telescope so that he could study the heavens. The moon was his first subject and Galileo was surprised to find that the moon was not the smooth sphere predicted by Aristotle but instead had a surface full of ridges and craters. His observations also enabled him to dispel the notion that the moon emits light; Galileo determined that the luminosity of the moon is based solely upon sunlight reflected off its surface. Galileo also studied the hazy band of stars known as the Milky Way and found that it was an immense collection of stars that he speculated were too numerous to count. He also discovered that Venus has phases like those of the moon.

Fortified by the successes of his investigations, Galileo turned his attention to Jupiter and discovered several of its moons. This discovery, more than any of the other ones made by Galileo, was the one that created the most excitement because it appeared to be a miniature version of the solar system with the massive planet Jupiter being circled by several tiny moons. What made the discovery of Jupiter's moons particularly fortuitous was that it offered a live model of the Copernican system that could easily serve as an analogy to the solar system as a whole. In any event, Galileo was prompted to publish *The Starry Messenger* in 1610 in which he recorded his many observations in a straightforward, factual manner. The book was a great commercial success and, somewhat surprisingly, did not elicit a hostile response from the church. Less profitable was Galileo's ambitious attempt to sell titles to the stars to various wealthy individuals in Italy.

But it was Galileo's experiments with motion that helped him describe mathematically the dynamics of the universe and invent the discipline of classical mechanics. In his masterpiece *Dialogues on the Two New Sciences*, which was published in 1638, Galileo offered his views of motion, gravitation, and acceleration; he used his experimental results to demolish both the Aristotelian philosophy and the Ptolemaic cosmology that had stifled scientific inquiry for two thousand years. Instead he extolled the virtues of both the empirical approach to science and the heliocentric theory itself. But as illustrated by the controversy that occurred following the 1632 publication of his *Dialogues Concerning the Two Chief Systems of the World*, not everyone wanted to see Galileo attack Ptolemy's geocentric system. Indeed, the church, which had a great deal of its own prestige and authority vested in both Aristotelianism and Ptolemy's system, viewed the publication as a direct challenge to both its primacy in spiritual affairs as well as its leading role in defining the importance of the earth in the universe. This publication led the church to summon Galileo to appear before the Inquisition. Faced with the prospect of being sent to prison for life or even tortured or killed, Galileo recanted his views. He was confined to a palace owned by a friend, where he spent much of the rest of his life in virtual seclusion. However, he took advantage of the opportunity to write his *Two Sciences* book and continue refining his views on motion and gravity. Soon thereafter, however, Galileo lost his eyesight, thereby putting an end to much of his work. To add further insult to injury, he was forced to endure constant surveillance by the church authorities until the time of his death in 1642.

Galileo's spirited investigations were motivated by his belief that all of the myriad of events in nature could be expressed using a comparatively small number of mathematical equations describing the most basic physical laws in nature. Of course Galileo lived in a time in which there were not very many such equations actually known to humanity. Thus his assumptions were as much an article of faith as a conviction borne of any real awareness of the great potential offered by mathematically based scientific investigations. But he believed that he could use deductive reasoning and proceed from basic assumptions to general conclusions to construct an entire scientific methodology that would gradually uncover the mysteries of the world. Now one can only imagine what Galileo's contemporaries must have thought when he began to talk about the motions of bodies in space (such as the earth) or the orbital paths of the planets around the sun. He might have been viewed as something of an eccentric. But Galileo was not one to worry about his popularity and, being a religious man, believed that he was doing God's work in seeking to discover the basic laws of nature. He certainly did not view his investigations as an affront to religion but more as a glorification of what he considered to be the Creator's universe.

Galileo's investigations of nature centered upon the phenomena of moving bodies which caused him to question the basic Aristotelian ideas about motion. Galileo, being an intellectual skeptic, immediately questioned Aristotle's idea that any body would remain at a state of rest unless acted upon by some force. Aristotle's view had a certain undeniable appeal and had dominated scientific thinking for nearly two millennia. Although Aristotelianism had taken root almost everywhere, manifesting in university lecture halls as well as church pronouncements, Galileo believed that one had to consider motion in a somewhat more sophisticated way. If one ignored the resistance encountered by an object moving along a surface, then a moving object should continue endlessly onward in a straight line. Here, Galileo had to rely on his own deductions and thought experiments because he could not create a frictionless surface; the then unknown force of gravity also made it impossible for Galileo to duplicate his thought experiments. But he was not hesitant to follow his basic deductions about moving objects to their natural conclusions and thus was able to offer a more complex explanation of motion that managed to infuriate the clergy in Europe. If Galileo's desire was to annoy as many people as possible in the shortest amount of time, then his dismantling of Aristotelian physics offered the perfect opportunity. As Galileo would later discover, it almost earned him a date to be burned at the stake, but as we already noted, he was practical enough to disavow his own revolutionary ideas at sword point in order to save himself and continue the fight until another day. Galileo could see little point of risking the wrath of the church and becoming a martyr for the new science—particularly when he was not sure that the new science, primitive as it was, would survive beyond his own lifetime.

Galileo's principle of motion posited that an object at rest would continue to remain at rest and an object in motion would continue to remain in motion at a constant speed when moving along a frictionless surface, unless an outside force intervened. But Galileo realized that the intervening force would have to be sufficiently large in order to overcome the object's resistance to a change in velocity. (Physicists use the term *velocity* to refer to the speed of an object moving in a particular direction.) If the object was the size of a boulder and the intervening force was a housefly that happened to fly in the way of the boulder, then it did not take an Einstein to figure out that the direction and velocity of the object (in this case, the boulder) would remain virtually unchanged. But if Galileo could toss a few other rocks in the way, then the mass of these objects might be sufficiently great to overcome the inertial resistance of the boulder.

Galileo then wondered how a moving body would be affected if a force was applied to it. A cannonball, for example, would be slowed down due to the air resistance and, ultimately, the unfortunate soldier on whom it landed. In the absence of such resistance, the cannonball would continue to move onward forever (if we discount the effect of gravitation—a phenomenon, as noted, that had not yet been discovered in Galileo's time). By the same token, that cannonball would continue to accelerate in velocity without limit if force were continuously applied to it. These ponderings led Galileo to what would become known as the second law of motion; it equates force to the product of a body's mass and its acceleration. This formula would later be expressed algebraically by Isaac Newton as $F = ma$, where F is the force, m is the mass of the body, and a is the acceleration. This formula was one of the cornerstones of classical physics. To earlier generations, it was as well known as Einstein's mass-energy equivalence equation ($E = mc^2$) is today. In any event, this equation was universal in its application to moving bodies, regardless of whether they were accelerating or slowing down.

Galileo's detective work led him to conclude that all objects in a vacuum fall to the earth at the same rate of acceleration—32 feet per second—and this can be expressed algebraically as $v = 32t$. So one can determine the rate of acceleration of an object at any point in time in its journey merely by plugging in the values for time t. After 10 seconds, for example, the rate of acceleration would be 320 feet per second. We could test this equation ourselves by tossing weights off towers of various heights and tracking their rates of acceleration. After a number of trials we would find, much to our relief, that the rates of velocity did correspond with those predicted by the equation. We would not expect the observed values to be exactly equal to those predicted by the equation because some of our falling weights might be caught in updrafts and others might hit passing flocks of birds. Even though we now know about air resistance and other extraneous factors, we would still find Galileo's equation to be verified.

Galileo's work in mathematics and science was motivated by his underlying desire to find universal patterns in nature. The equation $F = ma$ (which would later be offered by Newton) equating force and the product of mass and acceleration was applicable on the earth or the most distant star. In the same way, the formula expressing the distance a body falls in a time t, $d = 16t^2$, is applicable throughout the universe and will yield the same results whether we use information from Pisa, Italy, or from the surface of the moon. Because of the need to assume that such things as air resistance and friction do not exist, these equations expressed relationships regardless of the objects involved. As a result, Galileo's formula relating distance and time would apply to an ingot of gold or a piece of paper. If Galileo were to drop both gold and paper off of the tower of Pisa, we might find that the gold would plummet to the ground whereas the paper would gently float and waft downward until landing on the ground. But Galileo's formula tells us that if we could discount the air resistance and friction and, in effect, create a vacuum, then both the gold and the paper would land on the ground at the same time. This conclusion is very puzzling to most people who have great difficulty imagining two such different things falling to the ground at the same time. But this process of idealizing the environments in which such experiments occur is a necessary part of scientific inquiry because it makes it possible to simplify these experiments by ignoring extraneous matters (e.g., air resistance). Otherwise, the process of scientific investigation would be much more complicated and greatly hampered by the need to take into account every variable that arises in the real world.

Galileo also made important discoveries regarding the paths traveled by falling objects that were subject to different types of forces. Suppose Galileo was standing at the top of the Tower of Pisa and tossed a coin off the top. The coin would be subject to both the horizontal impetus of Galileo's push as well as its own downward vertical acceleration as it fell to earth. But the path followed by the plummeting coin would not be a straight line; it would instead be a curve similar to that of a parabola. If the coin was not too massive, Galileo's throw might cause it to go further outward in a horizontal direction. So the coin would fall along a path that continued to curve slightly away from the tower until it hit the ground. Being a curious fellow, Galileo might wonder about the paths that heavier falling objects like anvils would follow. Yet an anvil would still be thrown outward from the tower by Galileo's push. But because the anvil is a much more massive object than the coin, its horizontal displacement would be fairly small and it would fall to earth in a path that, while having some curvature to it, would more closely approximate a straight line than the path followed by the coin. The simple reason for this difference is that Galileo would not be able to toss the anvil outward as far as he could the coin. Thus the anvil would not be thrust outward as far from the tower before it began falling back to earth. But even though the anvil fell downward in a path more closely approximating a line running parallel with the wall of the tower of Pisa, it would not hit the ground any faster than the coin. Now this may seem counterintuitive because the curve of the coin's path to the ground would arc further outward than that of the anvil.

But the fact of the matter is that both the coin and the anvil still fall the same vertical distance from the top of the tower to the ground. The fact that the coin may have a greater amount of horizontal displacement when it gets tossed out of the tower than the anvil does not affect the velocity at which both will plummet to the earth.

Galileo continues to rank as one of the most important figures in the history of science and can be considered as perhaps the most influential voice in the shaping of our modern method of scientific investigation. His work owed a great philosophical debt to Euclid, whose own work in geometry inspired Galileo to try to isolate a small number of fundamental laws of nature from which many important theoretical discoveries could be made. But Galileo was also a first-rate experimentalist who was not afraid to get his hands dirty or even climb up to the top of a tower and toss objects over the side. But his practicality also dictated a strong sense of self-preservation that led him to choose his battles with the church authorities, even though he was unable to avoid being placed under house arrest for several years. Galileo showed great courage in almost single-handedly creating the modern scientific method and melding theory and experimentation. Because he was not able to enlist many allies in his intellectual battle against Rome, much of his spirited legacy was allowed to stagnate for several decades following his death in 1642.

fourteen

ISAAC NEWTON AND THE SEARCH FOR UNIVERSAL LAWS

As the seventeenth century unfolded, those few Europeans who thought about such esoteric matters as the orbits of the planets and the rates at which falling objects would hit the ground would have felt a sense of pride and wonder. By the time the Italian astronomer Galileo had died in 1642, the intellectual landscape of Europe had been shaken to its very core. Copernicus had offered a revolutionary view of the cosmos that uprooted the earth from its exalted position at the center of the universe to a point of comparative obscurity circling around the sun with several companion planets. Kepler had derived the first mathematical relationship that related the distance of the planet from the sun to the amount of time it took to complete its orbit. Descartes broke new ground, uniting algebra and coordinate geometry. Last but not least, Galileo had developed the basic laws of classical mechanics and, perhaps most importantly, defined the parameters of the modern scientific method. So it would not be surprising to find that the intellectually curious would feel that they were well on their way to

239

becoming the undisputed masters of their domain by understanding the universe in a more fundamental way than had previously been thought possible.

But this confidence was premature because science as an enterprise in Europe was still in its infancy and many were reluctant to cast aside the baggage of astrology and mysticism that continued to slow its progress. Still, four brilliant persons—Copernicus, Kepler, Descartes, and Galileo—had taken very important steps that required both originality and courage. And a key element in this movement from the Aristotelian geocentric cosmos that stubbornly prevailed for two thousand years was the evolution toward a body of scientific knowledge grounded on quantitative reasoning. It was at this point in our story that the groundwork was laid for the work of Isaac Newton that would forever fuse mathematics and science together and change the ways in which we study and perceive our universe.

Newton's place in history is secure because he made perhaps the greatest contributions of any single individual to science and mathematics through his formulation of the law of gravitation, his laws of motion, his theory of light, and his invention of the calculus. Although Newton certainly benefited from the work of his predecessors such as Kepler and Galileo, he saw further than anyone before him and created a unified system for understanding the motions of all objects—great and small—in the universe. What makes his discoveries and inventions all the more incredible is that he achieved all of them in the space of about two years. He made them as a young man living on the family farm after having fled an outbreak of the plague at Cambridge University.

Isaac Newton (1642–1727) was born at Woolsthorpe, Lincolnshire, England. He was so puny and sickly as a baby that the family made tentative plans for the funeral, never expecting that he would somehow manage to survive, let alone thrive. As a young child, Newton never displayed any visible signs of genius. Indeed, he was absent-minded and often neglected his studies. In today's parlance, he would have been considered "slow" and possibly steered toward a vocational career. He enjoyed building devices such as windmills and water clocks and tinkering with mechanical objects. His days on the family farm were typical of rural life at that time but he did attend several schools. It was only after his father died that the teenage Isaac left school to help his mother manage the farm. But Isaac was not a farmer and, despite having spent his entire life around agriculture and livestock, was quite inept at farming the land. Moreover, he did not like to tend the cows and sheep and would sometimes forget to bring them back in from the pastures. Indeed, it was with some relief that the family learned of Isaac's admission to Trinity College at Cambridge University as it meant that he would have to end his career as a farmer.

Although one might think that Newton's genius manifested itself while attending Cambridge University, it is more accurate to say that Newton did not stand out apart from his contemporaries at the university. He was not viewed as being particularly bright by his professors and did not receive any honors when he graduated in 1665 from Cambridge. His return to the farm was not greeted with jubilation by his family who remembered all too well his disastrous experiences as a farmer. But it soon became clear that Newton was more interested in pursuing his own studies than herding sheep. Over the course of the next two years Newton would have the leisure time to think about questions relating to the universe that had intrigued him for years. What his family thought of Newton's postgraduate career as a "thinker" is unclear. Subsequent events would show that his was an intellect that burst forth in an unprecedented explosion of creativity. Yet the catalyst for these discoveries remains unknown because Newton did his great work alone. He did not even bother to publish the results of his investigations until some two decades later and only after the insistence of his good friend, the English astronomer Sir Edmund Halley.

But whatever factors played a role in Newton's creative outburst and resulted in his discoveries, it is undeniable that his knowledge of the earlier discoveries of Galileo, Kepler, and Descartes converged with his own unparalleled intuitive insights into nature and mathematics. They resulted in his discovery of the composition of light, his invention of the calculus (which was arrived at independently by the German philosopher Gottfried Wilhelm von Leibnitz), his laws of motion, and his formulation of the universal law of gravitation. Newton recast virtually the entire body of scientific knowledge into a new mathematical framework and demonstrated that the laws of physics were applicable throughout the universe.

Despite these achievements, Newton said very little about his own discoveries during his self-imposed exile and quietly returned to Cambridge University after two years to complete the requirements for his master's degree and to join the faculty. Newton impressed the faculty members enough to be appointed Lucasian professor of mathematics following the resignation of his teacher, Isaac Barrow. But Newton's appointment was due less to the faculty's recognition of his as-of-yet undiscovered talents than it was the result of Barrow's vigorous lobbying of his colleagues for Newton's appointment. Having no publications to his credit and having said little about his own work to many scientists in England, Newton's appointment at the age of twenty-six must have seemed to many to be more a favor to Barrow than a coup involving the hiring of an unparalleled scientific talent.

As a professor, Newton was now responsible for giving lectures to students. But he was not the most gifted of orators, being very aloof and unable or unwilling to engage his students in meaningful dialogues. Indeed, very few students bothered to attend Newton's lectures even though he would occasionally present information about some of his own discoveries—most of which was ignored by them as well as the other members of the faculty.

Newton's first venture into the world of academic publishing was a paper he wrote suggesting that the composition of light was corpuscular (particlelike) in nature. Because Newton's theory of light challenged many of the assumptions underlying the wave theory of light that was then accepted by most scientists, Newton's work was criticized, sometimes unfairly, by some of his colleagues. But much of the criticisms, even at this early stage in Newton's career, were along nationalistic lines. Although they would eventually be universally acclaimed, Newton's discoveries were initially resisted by many of the scientists on the European continent, whereas much of his early support came from his fellow scientists in Britain. His theory of light, which assumed that light consists of particles, challenged a theory of light offered by the Dutch scientist Huygens that presented light as being undulatory (wavelike) in nature. Similarly, great controversies over the credit that should be given for the discovery of the calculus arose because Newton delayed publishing his discovery while the German von Leibnitz discovered the calculus in the intervening two decades. Nowadays, most historians credit Newton with having made the initial discovery but acknowledge that Leibnitz offered certain improvements, including a superior form of notation that continues to be used to the present day.

So Newton, whose greatest pleasure was intellectual research, learned that the scientific world was not simply going to recognize and applaud his achievements because science, like all fields of human endeavor, is ultimately one with political elements. Just having a well-reasoned theory that can be verified is not always enough because the prevailing scientific orthodoxy may not be receptive to a radical new idea—no matter how logical or coherent it may be. And many people would be surprised to find that this reluctance to embrace to new ideas extends to the community of mathematicians who, time and time again, have not initially accepted ideas that did not fit neatly within the dominant paradigms of the time. Still, Newton's invention of the calculus did not encounter the hostility of some more recent innovations, such as the nineteenth-century German mathematician Georg Cantor's set theory, because the scientific community had already been informed of the work of Leibnitz (and had gradually accepted the worth of the calculus by the time Newton formally announced his own discoveries in mathematics).

But Newton's reluctance to publish his results and become embroiled in controversies might have caused him not to reveal his discoveries at all had it not been for the repeated urgings of Halley (he was the famous British astronomer who discovered the comet that bears his name and passes by the earth every seventy-six years). Halley's role in the public dissemination of Newton's work was invaluable. He not only served as the cheerleader who convinced the reluctant Newton to summarize his research but also edited the proofs and subsidized the publication itself. The work, entitled *Mathematical Principles of Natural Philosophy*, was finally published in 1687. It laid out in an excruciatingly detailed axiomatic narrative the foundations of what would become known as Newtonian physics. The work, originally published in Latin, was made even more complicated by the fact that Newton had used geometric proofs to substantiate his findings, instead of the much more elegant calculus that would have greatly simplified the presentation of the work. But the choices were not accidental; Newton wanted to ground his discoveries, to the greatest extent possible, in the legitimacy offered by classical geometry and the use of Latin. Because the work was not easy reading, the reaction of the scientific world was not swift. Indeed, there were very few people in the world at that time who could understand Newton's work in all of its complexity. Some of Newton's detractors were not altogether sure if Newton himself was cognizant of the ramifications of his work. But those few who managed to wade through Newton's tedious proofs and turgid prose were rewarded with a dramatically different view of the universe in which all forces were mathematical in origin and governed by certain immutable physical laws. Newton's *Principles* gave rise to the metaphor of the watchspring universe in which all creation was imagined to be a gigantic device wound up by the hand of God and left to run on its own, guided only by the laws of classical mechanics. As such, Newton's vision was one of a universe in which God was separate and aloof, not involved in the minute details of everyday life. Needless to say, this paradigm had dramatic philosophical ramifications for such concepts as determinism, particularly as it seemed to leave humanity separate in certain ways from a benevolent Creator who watched the affairs of humanity with amused detachment from afar.

Newton's *Principles,* despite its difficult format, was imbued with the Galilean spirit; it envisioned science as an enterprise that should explain the manifestations of nature in precise mathematical laws. Although this was sort of a philosophy of science, Newton, like Galileo before him, was reluctant to speculate as to the "why" questions underlying his three laws of motion or his law of gravitation. Instead he was concerned with describing the ways in which these laws could be defined and verified. He did not want to waste time and energy in needless arguments regarding the reason the universe was created or speculate about the process that causes objects to be gravitationally attracted to each other. Instead his *Principles* provided a framework for his laws of motion and gravitation, and brought an unprecedented degree of mathematical order to the study of physics. This extensive use of mathematics also foreshadowed the modern theoretical approach in which the mathematical equation would play a critical role in both the formulation and verification of scientific theories.

Newton's investigation of gravitation was prompted by his desire to unify Kepler's laws of planetary motion and Galileo's laws of terrestrial motion. He assumed that the two had to be inextricably linked in some way and, taking a cue from some of the scientists of his day, thought that the answer would lie in the mutual gravitational attraction that seemed to exist among all objects. Newton saw that Galileo's law of motion would predict that the earth would move in a straight line through space and not in an elliptical path around the sun as demonstrated by Kepler. As no one could ever find a very long string connecting the earth to the sun, there seemed to be no other explanation except one suggesting that both the earth and the sun are gravitationally attracted to each other and that this mutual attraction causes the earth to fall continually inward toward the sun. Fortunately for Newton and all humanity, however, this inward fall is offset by the centrifugal force of the earth moving around the sun that counterbalances the attractive effect of the gravitational force. The two forces thus strike a happy medium between them so that the earth avoids sailing off into the cold void of space or being incinerated by the sun.

This sounded like a very cogent idea, but no one knew how to tie the laws of Kepler and Galileo together. Newton believed that there had to be a single equation or set of equations that would apply to both sets of laws and show that both sets of laws actually describe different aspects of the same general force—gravitation. To try to quantify this gravitational attraction, Newton turned his attention to the orbital path of the moon around the sun. His great insight was the realization that he could calculate a value to quantify the earth's gravitational pull on the moon. Newton picked an arbitrary point on the orbit of the moon; he then imagined how far outward the moon would continue to travel in a straight line if the gravitational force of the earth ceased to exist for one second as the moon moved from one point to another. He then drew a second line back to the orbital path, which represented the distance that the moon was accelerated downward to the earth in that same second. By making a series of calculations, Newton concluded that an object on the surface of the earth would be accelerated downward at a rate of sixteen feet in the initial second of time. Newton's investigations gradually led him to conclude that there was some type of relationship between the intensity of the attractive forces between the moon and the earth (indeed, any two bodies) and the distances between their centers. But Newton was not the type of man who was content with shadowy or implicit relationships; he demanded precision because he realized that approximations did not quite cut it if he wished to construct an enduring physical theory. So he continued his work and was able to derive an equation to express the relationship between the masses of any two bodies and their distances from each other: $F = kmM/r^2$ where F is the gravitational force, m is the mass of the smaller object (in this case, the moon), M is the mass of the larger object (in this example, the earth), k is a universal constant, and r is the distance between the center of the earth to the center of the moon. Although this formula does not look terribly ferocious, it is one of the most important equations in the history of physics because it can express the gravitational attraction between any two like or unlike objects—whether they are moons, planets, or stars.

The law of gravitation holds that every object in the universe attracts every other object—no matter how imperceptibly—because every object has mass, and mass gives rise to gravitation. Newton's law could be used to calculate the gravitational force that is exerted by the earth upon your body, the ice cream truck driving down the street, or even the plane flying in the sky. The important point is that Newton's law is universal in its scope, and no object, regardless of how big or small it may be is exempt from its reach. However, the law of gravitation cannot be used as a panacea for all calculations: Gravitation is the most powerful force in the universe of planets, stars, and galaxies, but it is the also weakest force in the world of subatomic particles and thus cannot be used as the sole guide for understanding the interactions of protons, neutrons, electrons, and their constituent particles. However, Newton lived in a time in which the bubonic plague dominated peoples' thoughts. Indeed, Newton himself had seen people die from the periodic outbreaks of plague that would periodically sweep across England. More to the point, the technology of Newton's time was not very advanced. Scientists knew nothing of electron microscopes or particle accelerators and so there was no way to detect atoms; they remained a legacy of the Greek scientist Democritus—more a figment of the imagination than anything that could be detected by the senses.

This digression might seem very nit-picky because the law of gravitation was a supreme intellectual achievement both in the grandeur of its applicability and its elegant simplicity. It offered a perspective that compelled those who studied the philosophical implications of Newton's theory to view the universe as a unified system governed by the same physical laws that could be expressed in straightforward mathematical equations. But his conclusions were not at first embraced by the scientific community, due both to the extreme difficulty of wading through the detailed proofs of the *Principles* and also the differing philosophical and scientific perspectives of his colleagues—both in England and on the Continent. Continuing disputes over authorship added to the controversy over the value of Newtonian physics. Indeed, Newton's publication of his *Principles* was prompted in part by a dispute with Robert Hooke, the English experimental physicist, over who should receive credit for the discovery of the inverse square law and the law of gravitation.

Although Newton was ultimately credited with these discoveries due to his ability to describe them using mathematical equations, many historians of science including the British scientist J. D. Bernal in his *Science in History*[1] believe that many of the basic ideas underlying Newtonian physics—particularly the law of gravitation—originated with Hooke. The problem that arises in trying to attribute priority to one person or another is that it is very easy to propose the idea that all matter is gravitationally attracted to all other matter and thereby give people something interesting to ponder. But it is quite another thing to construct a mathematical framework and provide precise equations whereby the manifestations of this attractive force can be studied, quantified, and used to investigate the motions of objects—whether they are planets orbiting around the sun or debris being tossed off the top of the Tower of Pisa. Indeed, Newton himself acknowledged his intellectual debt to his predecessors, particularly Galileo and Kepler, in saying that his unparalleled scientific achievements had been made possible by his standing on the shoulders of giants. By the same token, however, Newton did not take kindly to the criticisms or conflicting claims of priority made by contemporaries such as Hooke, a man who was one of the foremost experimental physicists of his time, but one whose investigations were greatly limited by his poor mathematical skills. Hooke was also something of a publicist's nightmare, being physically unappealing and cantankerous, so much so that even his admirers had difficulty defending his periodic outbursts. But he had a fertile mind and was not afraid to wander through vast reaches of the scientific realm, making important contributions to astronomy, mechanics, and even biology. He was also the first curator of the Royal Society, an honor more in name than in renumeration, but one that was a testament to Hooke's facility with scientific instruments. In any event, Hooke should have been given greater credit by posterity for his intuitive flashes of brilliance. These accolades should be justifiably tempered, however, due to Hooke's inability to apply his ideas about gravitation in a mathematically significant way.

Newton's efforts to define precisely the force of attraction between any two objects led him to formulate the following equation: $F = kMm/r^2$, where F is the force between the two objects M and m, r is the distance between the two objects and k is a universal constant.[2] In his initial investigations, Newton used the earth and the moon as his two objects. He calculated the attractive force exerted by the earth upon the moon and ultimately derived a value for the force by which the moon is pulled toward the earth. He thereby substantiated his suspicion that all objects exert an attractive force upon each other. But Newton, being a first-rate mathematician, was not content to limit himself to studying the motions of the moon by itself, and expanded his studies to include the earth's attraction upon objects located at its surface. This equation could thus be used to calculate the force between the earth and a mill or a horse or any other object on the surface of the planet.

Newton's equation for the gravitational attraction between two objects could be manipulated to provide valuable information about what might appear to be completely unrelated phenomena. It could even be used to tell us how to calculate the mass of the earth. How might this be done? One could begin with the formula for Newton's second law of motion, $F = ma$, which states that the force F acting on a body of mass m is equal to the mass of that body multiplied by its acceleration a. One could then set up the following equation: $F = ma = kMm/r^2$, which could be utilized to calculate the acceleration experienced by a freely falling body. Newton himself merely took the equation $ma = kMm/r^2$ and then divided both sides by m to get the equation $a = kM/r^2$. For someone as intuitively brilliant as Newton, it was only a matter of substituting values for k, M, and r to get the value expressing the acceleration of an object toward the earth, which Newton, after some deliberation, determined was equal to 32 feet per second. Once the English physicist Henry Cavendish calculated the value for the constant k in Newton's formula for universal gravitation, it was possible to calculate the mass of the earth because all of the other variables in the equation were now known. Sadly for Newton, Cavendish made his discovery almost a century after Newton discovered the law of gravitation and so Newton was not able to benefit from Cavendish's work. The constant was small enough, however, that it could be neglected for many types of calculations. The real excitement of Cavendish's achievement came once scientists realized that all of the variables of Newton's equation—save for the mass M of the planet earth—could be quantified. Consequently, one could manipulate the equation so that the mass of the earth could be calculated. But knowing the mass of the earth does have real meaning to modern physicists who must consider the effects of gravity in many matters ranging from the evolution of stars to the fate of the universe itself. Now Newton was certainly not expecting to predict the fate of the universe from his law of universal gravitation because he believed that the universe was essentially static and that its myriad of stars were scattered throughout space, unchanging through time. But he could see how knowledge of the different variables in his formula could yield new information, merely by his restructuring the equation. Thus additional information could be obtained for the particular value of the single variable placed by itself on one side of the equation—whether it be the acceleration of gravity, the mass of the earth, or even the constant of universal gravitation. This use of mathematics should be of particular interest to us because we have seen how the ability to manipulate equations has been of inestimable value to scientists in learning how to quantify nature.

Newton was more than just another scientist who had a knack for stumbling across fundamental laws of nature. He is also one of the very few persons in history to have made monumental contributions to both science and mathematics. Up to this point, however, we have talked more about his scientific work than his mathematical achievements. But his work in the calculus, as much as that in physics, ranks among the greatest achievements in the history of humanity. But his mathematical contributions—just like those he made in the physical sciences—were not made in a vacuum. Indeed, Newton's greatest gift may have been his ability to bring together disparate intellectual threads and weave them together into seamless tapestries of mathematically based theory.

The Calculus

Newton's discovery of the calculus—the mathematics of the rate of change of variables in minute periods of time—did not occur one day as he thought "Maybe I will do something really important today, like discover the calculus." (He actually called it the "method of fluxions.") Newton's discovery of the calculus (or synthesis, as less charitably inclined Continental scientists who favored the claims of priority made by the German von Leibnitz were inclined to say) was made during his youthful exile at Woolsthorpe. Newton greatly benefited from the work of ancient mathematicians such as Archimedes as well as his more modern predecessors such as Descartes and Fermat. But it was Newton—more than any other figure in history—who brought the calculus to bear to solve innumerable mathematical problems that could not have otherwise easily been solved by the scientists of that day.

The calculus enjoys a certain mystique because it is not taken by most students who study mathematics in high school. It is particularly unfortunate because the calculus is far simpler and easier to master than algebra. Moreover, the calculus is fundamentally concerned with distinguishing between change and rates of change in variables. There is admittedly more to the calculus than just these two concepts, but we do need to know something about both to solve the myriad of technical problems that have yielded to the calculus and made possible our modern industrial civilization.

The terms "change" and "rates of change" sound abstract but they can be easily understood by using the very familiar example of a country drive in the family car. After all, change is a straightforward concept. In setting out on our drive, our speed changes as we start from our house and then drive down the street to the expressway. After three hours we might find that we had driven 150 miles at an average speed of 50 miles per hour. This is not the type of performance that will open the doors to a professional racing career but it does illustrate the concept of driving at an average speed of 50 miles per hour. A change in speed, by contrast, occurs when we accelerate the car from 30 to 60 miles per hour in order to get around a particularly slow driver who is plodding along the highway. Acceleration by its very nature is a change in speed; it is very different from driving along at a constant rate of speed such as our average speed of 50 miles per hour. Acceleration is dynamic and invigorating. Driving at a constant rate of speed, by contrast, does not excite the senses because its constancy does not cause us to feel the thrill of movement along the highway.

Newton took the concept of acceleration one step further and considered changes in speed at instantaneous intervals (that is, imperceptibly small units of time). This is not a purely theoretical argument because objects can undergo wrenching changes in speed in very short periods of time. If you jump off of a thirty-story building, you cannot only verify the acceleration of gravity on the way down, but you can also experience an instantaneous rate of change in your downward velocity when you hit the ground. You may not be very interested in your instantaneous rate of change after such a landing (assuming you actually regain consciousness) but it is nevertheless a very profound change in your rate of acceleration.

And while you might not want to know your speed at the very instant you hit the ground, you might want to know whether there is any real value in being able to calculate an instantaneous speed. The speed of a meteor falling toward earth, for example, calculated at any single instant in the course of its journey would be its instantaneous speed. Instantaneous speed was also a topic of real concern to Newton because Kepler's law of planetary motion states that the planets move around the sun at constantly varying speeds. As a result, any analysis of its average speed over the entire course of its orbit will not be very helpful in revealing the velocity of the planet at any given point.

To return to our country drive, we know that we might average fifty miles per hour over the course of three hours, but we do not know how fast we are going in the sixty-fifth minute, for example, of our journey based upon our knowledge of our average speed. To discuss the concept of instantaneous speed in a meaningful way, however, we need to shorten the time period from minutes to seconds to fractions of seconds. But we find that if we move to an instant in which the elapsed time is essentially zero, then it appears that we would be dealing with a meaningless quantity because we would be traveling zero distance in zero time.

But Newton was far too clever to get caught by such obvious traps; his genius was to define the instantaneous speed at any given point as the average speed of an object as the unit of time *approaches* zero. Here he discarded the notion of using the formula for average speed (distance traveled divided by time elapsed) and instead moved to the concept that there is a number that is approached by average speeds over increasingly smaller intervals of time. As the change in t becomes smaller and smaller (which physicists write as Δt), then the average speed of the object is calculated over a correspondingly shorter interval of time. Newton reasoned that the instantaneous speed of an object could only be obtained once Δt became zero. This insight was the basis for what would become known as the differential calculus and, indeed, much of analysis itself, which in turn gave rise to great advances in both mathematics and the sciences that have continued to the present day.

No doubt you are wondering whether this conversation is leading to the examination of a mathematical equation. You will be pleased to know that we can consider real-world applications of the concept of instantaneous speed by returning to our friend Galileo and the objects he might try to toss off of the tower. If we wanted to determine the instantaneous speed of a falling object exactly four seconds after it was shoved over the side, we would go back to Galileo's formula $d = 16t^2$ and, substituting 4 for t, would find the distance would be equal to $d = 256$. So we know that Galileo's object would fall 256 feet in 4 seconds. This may seem familiar because we have already talked about it in some detail in the previous chapter. But this expression is known as a differential equation because it provides us with information about the instantaneous rate of change of one variable with regard to another variable. This is also a functional relationship because some change in one variable, x, will cause a change in another, a change of y. Hence, y is said to be a function of x because y is dependent upon the changes in x taking place and may be written as $y = f(x)$. Here, the information is provided in the form of the distance fallen at any given instant of time t. Newton himself was able to deduce Kepler's laws through the use of a differential equation. But it was the fact that the calculus could be used to solve problems involving one-, two-, and even three-dimension coordinate systems as well as all types of rates of change in the physical universe that was to make it such a valuable tool to modern society. Even though Newton (and Leibnitz) performed an invaluable task in creating the calculus from the chaotic snippets that had been strewn for many centuries throughout history by mathematicians, their work was rooted in intuition and physical examples; they did not engage in purely intellectual considerations devoid of real-world examples. Indeed, the logical structure of the calculus was not completely ironed out during Newton's time. Thus it remained for successors such as Augustin-Louis Cauchy to offer more rigorous proofs to substantiate or, in some cases, to shore up certain wobbly cornerstones in Newton's mathematical edifice. But the fact remains that the differential calculus is a very helpful intellectual tool that makes it possible for us to grapple with infinitesimal rates of change. It provides us with a technique for determining the rate of change of any relationship of one variable or quantity to another.

Given the respect with which the calculus is viewed today, you may be surprised to learn that many of the eighteenth- and nineteenth-century mathematicians, who used it so successfully to make important discoveries in mathematics and science, did not view it as a completely legitimate branch of mathematics. Much of this hostility sprang from the fact that Newton and Leibnitz had not focused primarily on the logical structure of the calculus—reasoning instead by both analogy and example. But these very same mathematicians who questioned the structural integrity of the calculus were also quick to recognize its value as a means for solving a host of complicated mathematical problems involving dynamic quantities. So the calculus was something of a "dirty secret" in that it got the job done but was not considered to be "first-rate" mathematics by certain of Newton's successors for many years after his death.

As for Newton himself, his discoveries brought him fame as the greatest scientist of all time, a view that many persons have continued to hold despite the phenomenal achievements made by Albert Einstein. The reasons offered by those persons ranking Newton above Einstein stem from the role he played in creating much of what would become the groundwork of modern physics while also devising a mathematical tool that could be used to buttress the theoretical predictions of his theories. Einstein, for his part, created his special and general theories of relativity and, as such, forced certain revisions in Newton's work. But much of Newtonian physics remains unchanged and students continue to be introduced to his famous equations that describe both his laws of motion and his law of universal gravitation.

Although Newton made other important contributions as he continued his career at Cambridge, we should also point out that his interests diverged into areas such as alchemy and astrology. Indeed, Newton spent a great deal of time in his later years authoring lengthy tracts on alchemy while also carrying on extensive correspondences regarding his own scientific work with many of the great minds of that era. Newton even dabbled in politics and was elected to parliament, but some say that his only comment during his stint as a politician during a session was to ask that a window be opened to let some fresh air into the room. He also served the government as Master of the Mint and oversaw the reissue of the nation's currency. As he grew older, Newton did not drop his scientific investigations entirely even though he showed progressively less interest in further scientific work. Newton undoubtedly maintained his knack for problem solving throughout his life and made some refinements in his theoretical work in both physics and mathematics. But it also seems clear that Newton's early genius, during his two-year stay at the family farm, was an unprecedented intellectual explosion that was rivaled only by Einstein's development of his theory of relativity as a young man at the beginning of the twentieth century while working in Switzerland as a patent examiner.

NEWTON'S MENTORS

Newton credited the grandeur of his works to several of his heroes such as Copernicus (who had revived the long-dormant theory of Aristarchus that the earth revolves around the sun); Kepler (who had discovered the three laws of planetary motion); and Galileo (who had largely invented the scientific method of forming and testing hypotheses and did extensive experiments in mechanics). This was clear in his remark that his work was possible only because he had stood on the shoulders of others. But the debt owed by Newton to these men may not have been so staggering as his comment might suggest. Copernicus had dared to resurrect a theory that had been dormant for nearly two millennia because of the apparent shortcomings of the clumsy geocentric (earth-centered universe) model offered by the Egyptian astronomer Ptolemy in the first century C.E. Although the Ptolemaic cosmos was quite ingenious and mathematically sophisticated for its time, it was unable to explain, as we have seen, certain apparent movements by some of the other visible planets of the solar system. Ptolemy had tried to correct these problems by introducing epicycles into the planetary orbits to explain these apparent retrograde motions. But Ptolemy's early successes in using epicycles to resolve some of these questions went to his head. His cosmological model soon became ladened with, and ultimately burdened by, an increasingly complicated array of epicycles that choked the beauty of the geocentric model and eventually led to its downfall. Despite its evident shortcomings, however, the geocentric model thrived for nearly 1,500 years until a young theology student named Nicolas Copernicus dared to consider that the earth might not occupy the favored position championed by the church. Copernicus risked incurring a death sentence from a reactionary papacy that had taken the geocentric model to heart because of its theological ramifications. (Why else would Jesus Christ have made his only appearance in the physical universe on a planet that was not at the center of the cosmos? After all, why would Jesus Christ have chosen to visit a minor planet that enjoyed no special position in the universe?) But Copernicus had not been a vocal crusader for his heliocentric theory, limiting himself during his lifetime to guarded discussions and correspondences with a few friends and acquaintances. Indeed, it was only near the end of his life that he arranged for the publication of his masterpiece *On the Revolutions of the Heavenly Spheres*, which outlined his criticisms of the unwieldy Ptolemaic model and suggested that the planets of the solar system revolve around the sun in circular orbits.

The German mathematician and astronomer Johannes Kepler was closer in spirit to Newton than was Copernicus. Whereas Copernicus had hypothesized a solar system of circular orbits due to his belief that such paths were the only appropriate perfect shapes for heavenly bodies, Kepler was not so easily swayed. Although he had initially assumed that the orbital paths of the planets are circular, Kepler became convinced, based upon his examinations of the astronomical data recorded by the great Danish astronomer Tycho Brahe, that the circular orbits had to be scrapped. Kepler seems to have been convinced that there was a simpler, more elegant explanation because his calculations showed that the Copernican model would still have to utilize Ptolemy's epicycles in order to match Brahe's tables. But as he proceeded to carry out laborious mind-numbing calculations and the weeks turned to months and the months to years, Kepler became more desperate to find a solution. His anxiety was made worse by the grinding poverty that had dogged him throughout his life, only being occasionally eased by clients who paid him to do astrological fortunetelling. Kepler was also badgered incessantly by his mother, a woman who had a vile temper and a seemingly interminable lifespan.

Kepler played with several different shapes such as ovals before stumbling upon the answer: the planets move in elliptical orbits around the sun. His discovery of the first laws of modern physics culminated with his third law; it states that the square of the orbital period of each planet is proportional to the cube of its mean distance from the sun. Kepler thus gave the world a single algebraic equation that described many of the dynamical features of the solar system. In doing so, he offered a glimpse of the pivotal role that mathematics could play in simplifying and, hence, unifying the seemingly chaotic face of the universe. It also established a pattern of deductive reasoning that would continue to be used by scientists to the present day in which physical laws would be derived from observations and expressed in mathematical form.

But Kepler had a darker, obsessive side to his personality. He was consumed with the idea that there was a relationship between the five known planets of the solar system and the five solids described by Plato (including the triangular pyramid and the cube), each of which could be inscribed in or superscribed along the surface of a sphere. Of course there was no physical significance to this presumed correspondence but the numerically obsessed Kepler was convinced that there was some sort of underlying pattern that would reveal itself to him if only he could derive the proper mathematics. Sadly, Kepler wasted much of his career in a frustrating and, ultimately, ill-fated attempt to prove that the solar system could be modeled based upon this presumed geometrical hierarchy.

The philosophical significance of Kepler's work was not lost upon Newton who recognized the profound power of mathematics to describe natural phenomena. Yet Newton was also interested in the practical applications of these laws and, therefore, became greatly enamored with the experimental work of the Italian scientist Galileo Galilei. Newton admired Galileo for his abilities as an instrument maker and his "roll-up-the-sleeves" approach to experimentation. Whereas Kepler would have tried to calculate the rates objects would fall from a tower using pen and paper, Galileo actually lugged weights up to the top of the tower and dropped them to the ground to see if there was any differences in the rates at which different objects (such as feathers and lead) fell. Galileo was also a role model for Newton because he had an almost arrogant confidence in his abilities as an experimentalist, ignoring the papal authorities who called for a slavish adherence to the geocentric model of the cosmos.

It was Galileo who had introduced mathematics into physics and done his best to purge the sciences of the suffocating superstitions that had been foisted upon it by a reactionary church. Galileo was perhaps even more important in Newton's intellectual development than Kepler because he formulated his law of freely falling bodies; it anticipated Newton's own laws of motion. Galileo argued that a body on a frictionless horizontal surface will continue to persist in its original state of motion or rest indefinitely. Taking this position carried certain political risks, however, because it flew in the face of the church-sanctioned Aristotelian philosophy that claimed all bodies eventually come to a state of rest if there is no force pushing them. Even though Aristotle had been dead for nearly two thousand years, his influence continued to shape (and, in many way, impede) the works of scientists and theologians alike. To challenge Aristotle was to risk excommunication by the church or more severe punishments such as being burned at the stake. But Galileo, to his credit, continued to carry out experiments and draw his own conclusions from his results in a manner not unlike that of the modern experimental scientists.

Galileo did extensive experiments with pendulums that led him to a primitive version of the law of conservation of energy, concluding that the bob of the pendulum would continue to swing without limit in a frictionless environment. Galileo even anticipated a sort of relativity theory, concluding from his experiments with falling weights that the laws of nature are invariant not only with moving and rotating objects but also when those objects change velocity. Indeed, Galileo's concept of relative motion was more advanced than the cosmos of absolute time and space that would be posited by Newton half a century later and would figure prominently in Einstein's own theory of special relativity.

But Galileo could claim a rare versatility that enabled him to move easily from the cerebral world of mechanistic theory to the practical considerations involved in constructing the first telescope.[3] When he turned his new instrument to the sky, the human perception of the observable universe was changed forever. Galileo himself was temporarily overcome by the spectacle of a celestial carousel previously unseen by mortal eyes. But he soon recovered his composure. He kept detailed notebooks of his discoveries that included many previously unseen stars and several moons of Jupiter and mountain ranges on the moon.

Galileo's pragmatism also extended to issues of self-preservation. When he was placed under house arrest for his heretical support of the Copernican cosmology, he decided that discretion was the better part of valor and recanted. But even as Galileo knelt, embracing the idea of a stationary earth with apparent enthusiasm, surrounded by his captors, he was likely muttering under his breath that the earth must move around the sun. Galileo's tactical maneuvers enabled him to avoid the fate that had befallen the Italian philosopher Giordano Bruno who had dared to propose that the universe consists of an infinity of worlds. Bruno had been engaged in a running battle with conservative clerical authorities for much of his adult life and had been repeatedly imprisoned for daring to question the primacy of the earth in the cosmos. The matter finally came to an end in the cavernous Campo dei Fiori in Rome where Bruno was burned at the stake on February 17, 1600. Bruno's death was not really a significant blow to science (except from Bruno's point of view) because Bruno's work, though offering a alternative vision of the universe, could boast little in the way of a coherent mathematical or logical foundation. Instead his infinity of worlds was based on his mystical (some would say hallucinogenic) view of the cosmos. It was not dependent on physical observation or described by mathematical equations. Yet it was remarkable because it did offer a much grander vision of the evening sky than most of the astronomers of that era could imagine. Bruno's untimely end was a convincing example of theological repudiation by the church in the finest tradition of Samuel Johnson's stone-kicking tantrum: "I refute Bruno thus."

Fortunately for Newton, seventeenth-century England was comparatively tolerant of new ideas. Most citizens jealously guarded their liberties and were wary of the ambitions of a church that they had seen as a threat to their nation since the time of Henry VIII. The heavy hand of religious orthodoxy was not as suffocating because the Church of England itself had been borne of a revolt against the papacy. Of course the revolt had been based more upon Henry VIII's inability to obtain Rome's approval of his decision to divorce his wife, Catherine of Aragon, and marry Anne Boleyn than any profound ideas of intellectual freedom, but the two purposes had still been served. Yet criticisms of the crown were not tolerated nor was England a bucolic paradise where young natural philosophers could leisurely stroll about thinking great thoughts. It was still ravaged by periodic outbreaks of deadly diseases including the bubonic plague. Indeed, it was a recurrence of the plague in 1665 that led to the closing of Cambridge University. Newton, then an undergraduate at Trinity College, retreated to the family farm at Lincolnshire where he spent much of the next two years puttering around the fields and thinking about the nature of space, time, and mathematics. Doubtless Newton's intellectual endeavors were aided not only by the absence of any discernible distractions (including members of the opposite sex, for whom Newton apparently had little interest) and an astounding ineptitude for animal husbandry, but also by what may have been an ability, perhaps never before or since equaled, to concentrate on a single problem with unrelenting intensity until it yielded an answer. Indeed, it was on this farm that Newton reportedly saw an apple fall out of a tree and wondered about the force that caused the fruit to drop to the earth. As he considered the matter further, he concluded that both the earth and the apple exert an attractive gravitational force upon each other. Once he reached this point, he then extended the concept of a pervasive gravitational field to all celestial bodies. Using his newly invented calculus, he calculated the forces that keep the planets in their orbits and showed how the motions of all the planets and stars in the universe can be explained in terms of his three laws of motion and his universal law of gravitation.

Newton thus laid much of the foundation for our modern mathematically based science. Unlike Galileo, who was obsessed with battling the pseudoscientific ideas of antiquity, Newton proposed a body of principles and proofs that gave natural philosophy (as physics was then called) a self-consistent, mathematically based logic that it retains to the present day. Newton's calculus enabled him to describe mathematically the motions of celestial bodies, and his laws of motion provided a dynamic model of the universe. Newton's supreme achievement in reasoning marked the culmination of what the American scholar John Herman Randall Jr. called "a complete mechanical interpretation of the world in exact, mathematical, deductive terms."[4] Indeed, Newton united the work of Kepler and Galileo "in one comprehensive set of principles, by calculating that the deflection of the moon from a straight path, that is, her fall towards the earth, exactly corresponded with the observed force of terrestrial gravitation; and he further showed that on his hypothesis Kepler's law of planetary motion followed mathematically from the law of gravitation."[5]

Newton's work offered profound physical insights because it demonstrated that the same laws of nature were invariant throughout the universe. It also obliterated any remaining thoughts that the laws of nature on earth might be somehow different due to it supposedly occupying a preferred (central) position in the cosmos. In a sense, Newton's work marked the first great unification theory because it offered quantifiable principles that governed the motions of the earth as well as those of the most distant stars. Newton resisted the obvious temptation to speculate about the cause of gravity but instead confined himself to positing that all bodies are endowed with the principle of universal gravitation.

Even at the young age of twenty-two, Newton showed a surprising intellectual maturity by searching for general principles and ignoring the more plodding inductive approach that had been used by many of his predecessors. Newton's boldness was aided by his intuitive grasp of mathematics and his ability to imagine how the motions of all the objects in the universe could be explained using a few simple mathematical formulas. Newton's success also inspired future generations of mathematicians to conclude (erroneously) that all of the events in the universe could be deduced mathematically from the principles of mechanics. The success of

Newton's theoretical edifice caused many to wonder whether science itself was at an end with only some minor skirmishes remaining to be fought in the battle for truth and knowledge.

Despite his remarkable feat of giving humanity a universe founded on logic, mathematics, and rational thought, Newton was extremely anxious that he be given the appropriate credit for his discoveries—even though he was reluctant to publish the results of his research. Indeed, he later became embroiled in several notoriously bitter controversies over credit for the discovery of the calculus and his theory of optics. Although no one had ever seriously contested Newton's central role in the creation of what would become known as Newtonian mechanics, Newton himself could not accept the idea that other persons such as the German scientist Gottfried Wilhelm von Leibnitz could have independently and contemporaneously discovered the same things (in Leibnitz's case, the calculus) without engaging in some sort of intellectual misappropriation. Even at this early stage in the development of modern science, there were enough talented persons working on mathematics and physical problems that some duplication of effort and simultaneous independent discovery was inevitable. But the dispute over the calculus mushroomed into an ongoing battle that became an issue of national honor with England's scientists pressing the cause of its favorite son and most of the Continental scientists siding with Leibnitz. In the end, both sides had to call it a draw and acknowledge Newton and Leibnitz as the coparents of the calculus.

Newton's own successes—in devising a mechanistic model of the universe in which the motions of the planets and stars could be explained using his laws of motion and his universal law of gravitation—represented (with the possible exception of Einstein's theory of relativity) the most stirring example of how a mathematical theory can impose a simple yet profound order upon an entire universe. No scientist works in a vacuum, however, and Newton was certainly no exception. He had benefited greatly from the earlier works of Copernicus, Kepler, and Galileo. But Newton, possessing perhaps the keenest intuition for analyzing the features of the physical universe, drew upon the sometimes disparate works of these individuals and almost single-handedly recast physics, astronomy, and philosophy in his own image.

But Newton's success depended on his ability to see patterns in nature where others saw only chaos. He was alone in being able to formulate a new world system that completely changed the way in which humanity viewed the universe. Yet Newton's greatest contribution to science may be one that is most overlooked by those who will readily crown him the preeminent scientist of all time: his rigorous and mathematically based reasoning that he used to formulate his mechanistic view of the universe. In short, Newton was inspired at some point in his early years, perhaps while he was still at Cambridge prior to the outbreak of the plague, to wonder about the dynamic forces in the world. He then sketched out his laws of motion and his law of universal gravitation, focusing in particular on calculating the rate at which an object is accelerated to earth. At the same time, he devised a completely new mathematical language (the calculus) to explain the mathematical relations among celestial objects. But Newton's true genius was to show mathematically that the same laws apply both to the motions of the planets as well as to the motions of objects on the earth, thus unifying terrestrial and celestial dynamics. His approach—for which scientists will forever owe him an incalculable debt—was to focus his masterwork, the *Principles of Natural Philosophy*, on describing the laws of motion and the law of gravity rather than trying to explain the causes of the physical processes he observed in nature. In short, Newton avoided the semantic quicksand inherent in speculating about "why" the universe is the way it is and instead concentrated his energies on describing it mathematically. Although criticized by some for refusing to delve into these deeper philosophical questions, Newton refused to become involved in what he viewed as intractable and unanswerable questions. He readily admitted the limitations of his approach in the following passage from the *Principia*:

[We] have explained the phenomena of the heavens and of our sea by the power of gravity, but have not yet assigned the cause of this power. . . . I have not been able to discover the cause of those properties of gravity from phenomena, and I frame no hypotheses; for whatever is not deduced from the phenomena is to be called an hypothesis; and hypotheses . . . have no place in experimental philosophy. In this philosophy particular propositions are inferred from the phenomena, and afterwards rendered general by induction. Thus it was that the impenetrability, the mobility, and the impulsive force of bodies, and the laws of motion and of gravitation, were discovered. And to us it is enough that gravity does really exist, and acts according to the laws which we have explained, and abundantly serves to account for all the motions of the celestial bodies, and of our sea.[6]

Newton's refusal to venture beyond observations and measurements and mathematics was fortunate, because it helped to map the terrain of knowledge that scientists needed to roam. They could leave behind religion and mysticism, and focus instead on that which they could readily observe and express in a quantifiable form. Of course this metamorphosis did not occur in one fell swoop and, indeed, had been proceeding in fits and starts before the time of Newton. But it was Newton's great work that drew science away from the shadows of Aristotelianism and religious orthodoxy forever, and gave it its own language and, indeed, its own identity. Critical to Newton's success was his investigative work that employed a process of deductive reasoning similar to what later would be called the modern scientific method.

THE SCIENTIFIC METHOD

What we refer to as the scientific method, as noted previously in this book, is arguably nothing more than what the American physicist Richard Feynman characterized as a process whereby "we try to put things together and try to understand this multitude of aspects [in nature] as perhaps resulting from the action of a relatively small number of elemental things and forces acting in an infinite variety of combinations."[7] According to Feynman, the scientific method was developed to make it possible to find simplicity amidst the plentitude of nature. It is a three-stage process consisting of observation, reason, and experiment that will, in Feynman's view, help us to learn the rules of the game played by nature. The challenges inherent in this task are likened by Feynman to a novice having to deduce the rules of the game of chess by watching others silently engage in play. But the rules to which Feynman refers are the basic laws of physics, which in turn require a thorough knowledge of mathematics. Feynman did not believe that we can necessarily learn the reasons that these rules exist but he did feel that science would gradually uncover most, if not all, of the rules themselves. But Feynman was quite concerned that our ability to understand the universe would be severely limited until such time as we were able to learn all of these rules—a process that might continue into the future indefinitely. He also wondered whether learning these rules would enable us to formulate theories that would explain all the phenomena of nature ranging from the most minute particles to the most massive supergalaxies.

The German physicist Max Born was also keenly aware of the importance of studying the physical world using an organized method of inquiry based on mathematics. But he realized that this method would not always enjoy unqualified succcess. Born believed the development of science to be "a picture of continuous and healthy growth, of unmistakable progress and construction, evident as much in its inward deepening as in its outward application to the technological mastery of Nature."[8] Yet Born was quick to point out that one also observes "at not infrequent intervals the occurrence of upheavals in the basic concepts of physics, actual revolutions in the world of ideas, whereby all our earlier knowledge seems to be swept away, and a new epoch of investigation to be inaugurated."[9] He would probably agree that science is at times driven by trends. Newton's theory of optics, for example, holds that light consists of a stream of particles. The Dutch physicist Christian Huygens, dissatisfied with the shortcomings of the corpuscular model, proposed that light consists of waves. Proponents of the two theories waged an ongoing battle. Huygens's theory gained the upper hand in the nineteenth century and Newton's theory returned the favor with the discovery of the photon at the start of the twentieth century. At the present time, though, the two camps agree that light has both corpuscular and undulatory characteristics. This debate illustrates the tendency of science to recycle itself over time because of the unwillingness of most scientists to accept an incomplete explanation as the definitive word on a given subject. This mutability of ideas about the universe may be frustrating to those who seek absolute certainty, but in Born's view, it is the fact that theories continually rise and fall that offers such appeal to the scientist.

Newton—along with Descartes, Kepler, Galileo, and Copernicus—was almost single-handedly responsible for creating the mathematically based, logically consistent science that we know today. Indeed, humanity owes an incredible debt of gratitude to these men because they risked professional ridicule and, in some cases, capital punishment to create an intellectual edifice that would impose an unyielding, immutable structure upon science. Were it not for their successful attempt to hoist science upon the skeleton of mathematics, our efforts to better understand the physical world would have continued to sputter because there would have been no universal yardstick such as that provided by mathematics whereby the most basic theories could be evaluated in quantitative terms. Indeed, a much more revolutionary change in the way we investigate and explore the universe was arguably wrought by these men than would take place even in the twentieth century with the discovery of the theory of relativity and quantum theory.

Unfortunately, the passage of time and the complacency that invariably sets in as revolutionary discoveries are assimilated has caused us to forget the significance of their works. Every generation has its own heroes and icons and is often equally quick to brush aside the standard-bearers of previous generations. So we do need to remind ourselves from time to time about men such as Newton, Galileo, Descartes, Kepler, and Copernicus who have been of supreme importance in the on-going effort to add to the reservoir of human knowledge and, indeed, our understanding of the world as a whole. Although our sciences probably would have still evolved if these men had never lived, it seems likely that it would have been a pale imitation of the beautiful theoretical edifice that we now have today.

afterword

My high school trigonometry teacher was a charming woman with a southern accent. She had a very unique attitude toward mathematics, viewing its mastery as a process in which "You've gotta suffer." Indeed, many of us can recall times in which we did not appreciate learning about mathematics, finding its nomenclature too difficult or its detailed proofs too confusing. But most of us can also recall feeling a sense of satisfaction when we mastered a particular concept or operation and were able to calculate something that had heretofore been a mystery. We then enjoyed a sense of accomplishment, having calculated the solution to an algebraic equation with several variables, or later, having solved a problem that was integral to the performance of our jobs.

A mastery of mathematics does require effort but my purpose in writing this book was to highlight some of the areas of mathematics that provide both intellectual stimulation and practical applications. It is my hope that the chapters of this book fulfill this criteria. Because this book is written at an introductory level, I have tried to avoid digressions into the more technical aspects of the subject matter; I did not want to lose sight of my purpose for writing—to entertain and to inform.

As work on the book proceeded, I realized that I would be remiss if I did not devote part of the book to the individuals most responsible for integrating mathematics with science. Consequently, I explored the most enduring contributions of Copernicus through Newton. Although some biographical information was provided about other individuals throughout the book such as Euclid, Pythagoras, and John Gaunt, the works of Newton and his predecessors represent the bridge between the mathematics of the past and the mathematically based science of the present. Their work, and the work of those who will follow, will continue to drive the development of mathematics. It is my desire that the reader will have enjoyed this excursion through some of the great concepts of our most profound thinkers whose ideas continue to play a significant role in our world.

notes

CHAPTER 1: THE ORIGINS OF MATHEMATICS

1. See Carl B. Boyer, *A History of Mathematics* (Princeton, N.J.: Princeton University Press, 1985), pp. 2–5 in which he offers a number a different ideas about the origin of counting in primitive societies, all of which seem to center upon the "contrast between one wolf and many, between one sheep and a herd, between one tree and a forest, suggests that one wolf, one sheep, and one tree have something in common—their uniqueness. In the same way it would be noticed that certain other groups, such as pairs, can be put into one-to-one correspondence" (p. 3).

2. Ibid.

3. See, for example, Levi Conant, *The Number Concept: Its Origin and Development* (New York, Macmillan, 1923).

4. See, for example, A. Seidenberg, "The Ritual Origin of Counting," *Archive for History of Exact Sciences* 2 (1962): 1–40.

5. Boyer, *A History of Mathematics*, p. 4.

6. According to Stephen F. Mason's *A History of the Sciences* (New York: Macmillan, 1962), "numbers provided a conceptual model of the universe, quantities and shapes determining the forms of all natural

objects." But this view of numbers was more than a purely quantitative notion: "At first they [the Pythagoreans] thought of numbers as geometrical, physical, and arithmetical entities made up of unit points or particles. . . . Thus, for the Pythagoreans, numbers had a geometrical shape as well as a quantitative size, and it was in this sense that they understood numbers to be the forms and images of natural objects," p. 29.

CHAPTER 3: REALLY BIG NUMBERS

1. A very entertaining book about the history and philosophy of the infinite can be found in Rudy Rucker's *Infinity and the Mind: The Science and Philosophy of the Infinite* (New York: Bantam, 1983).

2. Andrew Hodges, *Alan Turing: The Enigma* (New York: Walker & Company, 1983), p. 135.

3. Ibid.

4. Ibid., pp. 135–36.

CHAPTER 4: FRACTIONS

1. See generally W.W. Rouse Ball, *A Short Account of the History of Mathematics* (New York: Dover, 1960), pp. 3–5.

2. Morris Kline, *Mathematics in Western Culture* (Oxford: Oxford University Press, 1953), p. 16.

3. Ibid.

CHAPTER 5: ALGEBRA FOR EVERYONE

1. *The Universal Encyclopedia of Mathematics* (New York: Simon & Schuster, 1964), p. 65.

2. Ibid., p. 11.

CHAPTER 6: EUCLID'S MASTERPIECE

1. Ptolemy was one of Alexander the Great's generals who, following Alexander's death in 323 B.C.E., took over Egypt and ruled as king. He established the university and library at Alexandria and oversaw the development of Egypt as a great commercial power with several colonies throughout Asia Minor. He should not be confused with the great astronomer Claudius Ptolemy, who lived in Egypt in the first century C.E., and made important astronomical observations that led him to reject the idea that the earth moves around the sun. Instead he proposed in his work *The Almagest* that the earth is the center of the universe; a view that was largely accepted throughout Europe until the publication in 1543 by the Polish astronomer Nicolaus Copernicus of his theory that the sun—not the earth—is the center of the solar system.

2. W. W. Rouse Ball, *A Short Account of the History of Mathematics* (New York: Dover, 1960), p. 54.

3. Ibid.

4. Ibid., pp. 54–55.

5. Ibid., p. 54.

6. Ibid.

7. Bertrand Russell, *A History of Western Philosophy* (New York: Simon & Schuster, 1945), p. 211.

8. David E. Smith, *History of Mathematics* (New York: Dover, 1925), p. 274.

9. Ibid.

10. Ibid.

11. This is not to suggest that Aristotle's examination of geometry was superficial but merely that his inquiry did not entail the complexity or organizational skill of that shown by Euclid. Indeed, Bertrand Russell writes that Aristotle "is in many ways very different from all his predecessors. He is the first to write like a professor; his treatises are systematic, his discussions are divided into heads, he is a professional teacher, not an inspired prophet. His work is critical, careful, pedestrian, without any trace of Bacchic enthusiasm. The Orphic elements in Plato are watered down in Aristotle, and mixed with a strong dose of common sense; where he is Platonic, one feels that his natural temperament has been overpowered by the teaching to which he has been subjected. He is

not passionate, or in any profound sense religious." Russell, *A History of Western Philosophy*, p. 161.

12. Smith, *History of Mathematics*, p. 275.

13. See ibid., p. 276.

14. Ibid.

15. Ibid.

16. Ball, *A Short Account of the History of Mathematics*, p. 55.

17. Smith, *History of Mathematics*, p. 278.

18. Ibid.

CHAPTER 7: GEOMETRY AND MATHEMATICAL REASONING

1. David E. Smith, *History of Mathematics* (New York: Dover, 1958), p. 281.

2. Euclid's *Elements*, though often considered to be purely a compendium of geometrical knowledge, was actually a textbook of introductory mathematics. According to Carl B. Boyer, "[the *Elements* is] an introductory textbook covering all elementary mathematics—that is, arithmetic (in the sense of the English 'higher arithmetic' or the American 'theory of numbers'), synthetic geometry (of points, lines, planes, circles, and spheres) and algebra (not in the modern symbolic sense, but an equivalent in geometrical garb). It will be noted that the art of calculation is not included, for this was not a part of university instruction; nor was the study of the conics or higher plane curves part of the book, for these formed a part of more advanced mathematics." Carl B. Boyer, *A History of Mathematics* (Princeton, N.J.: Princeton University Press, 1985), p. 115.

3. Ibid., p. 116.

4. Ibid., pp. 116–17.

CHAPTER 8: ONE, TWO, THREE, TRIGONOMETRY

1. Morris Kline, *Mathematics in Western Culture* (Oxford: Oxford University Press, 1953), pp. 67–68.

2. Trigonometry has been defined as follows: "Trigonometry is the part of mathematics which has to do with the calculation of elements of

the triangle using the angular functions. At the basis of all such calculations is the right-angled triangle. All other triangles can be divided into two right-angled triangles by dropping an altitude from one vertex. Now if two right-angled triangles coincide in respect of one acute angle they are similar in shape; more precisely, the ratio of two corresponding sides in such triangles is the same. The ratio of two sides in a right-angled triangle is thus dependent only on the size of the acute angle and not on the lengths of the sides, i.e., it is a function of the angle (angular function)." *The Universal Encyclopedia of Mathematics* (New York: Simon & Schuster, 1964), pp. 470–71.

3. This example appeared in Morris Kline, *Mathematics and the Physical World* (New York: Dover, 1959), pp. 95–96.

4. Ibid., pp. 97–98.

CHAPTER 9: GRAPHS AND MORE GRAPHS

1. See Eric Temple Bell's *Men of Mathematics* (New York: Simon & Schuster, 1937), p. 98, in which the author defines variables and functions as follows: "A letter, say s, which can take on several different values during the course of a mathematical investigation is called a variable; for example s is a variable if it denotes the height of a falling body above the earth. . . . The word function or its Latin equivalent seems to have been introduced into mathematics by Leibnitz in 1694; the concept now dominates much of mathematics and is indispensable in science. Since Leibnitz' time the concept has been made precise. If y and x are two variables so related that whenever a numerical value is assigned to x there is determined a numerical value of y, then y is called a (one-valued, or uniform) function of x, and this is symbolized by writing y = f(x)."

CHAPTER 10: PROBABILITY AND GAMES OF CHANCE

1. A factorial is indicated by the "!" mark following a positive integer such as 1, 2, 3, 4, 5, and so on. When we show the expression 7!, for example, we are offering a short-hand way for expressing the following quantity: $7 \times 6 \times 5 \times 4 \times 3 \times 2 \times 1$, which is equal to 5040. *The Universal*

Encylcopedia of Mathematics (New York: Simon & Schuster, 1964), p. 219, provides the following formal definition for the term "factorial": "For positive integers *n* the product of the first *n* natural numbers is called '*n*-factorial' and is written $n! = 1 \cdot 2 \cdot 3 \ldots (n-1) \cdot n$. For n = 0, 0! Is defined as 1."

2. "The array of numbers in Pascal's triangle has the following properties: (1) Each number is the sum of the two numbers standing above it to the left and right; e.g., 10 = 4 + 6. (2) Each number is equal to the sum of all numbers in the left or right diagonal, beginning with the number immediately above to the left or right, and proceeding upwards; e.g., 15 = 5 + 4 + 3 + 2 + 1 and 15 = 10 + 4 + 1. (3) Each diagonal is an arithmetic sequence." See *The Universal Encyclopedia of Mathematics*, p. 323.

3. Brian Burrell, *Guide to Everyday Math* (Springfield, Mass.: Merriam-Webster, 1998), p. 204.

CHAPTER 12: COPERNICUS, KEPLER, AND THE RISE OF MATHEMATICS IN SCIENCE

1. Some might argue that Claudius Ptolemy's geocentric model was a notable—albeit flawed—attempt to fuse mathematics and astronomy. But Ptolemy's model was constructed to fit his preconceived views about the primacy of the earth in the solar system. As such, it was modified repeatedly due to the discrepancies between the circular orbits predicted by the model and the actual observations of the movements of the planets. As a result, it was more a purely mathematical model of the universe that was first and foremost a philosophical statement; the fact that there might be other explanations for the apparent motions of the planets such as a heliocentric solar system was not considered.

2. "Their contribution not only provided the most successful astronomical theory, demonstrated the power of mathematics to master nature's ways, and altered the intellectual life of Europe, but initiated a new pattern of scientific thinking in which mathematics was to play a far more fundamental role than it had in all the previous centuries." Morris Kline, *Mathematics and the Physical World* (New York: Dover, 1959), pp. 109–11.

3. Ibid., p. 112.

CHAPTER 13: DESCARTES, GALILEO, AND THE MATHEMATICAL UNIVERSE

1. In analytical geometry, "One field of mathematics is used to assist the development of another. As it was Descartes who first unified geometry and algebra in this way, the use of coordinates or sets of numbers to represent points and other geometrical structures is spoken of as Cartesian geometry, although Descartes himself did not use 'Cartesian coordinates.' We have an everyday use of coordinates in map-references and coordinates in analytical geometry are developed and refined from such examples as this. In the translation from geometry to algebra, those points which have a given property (which lie for example on a circle or sphere) translate into sets of coordinates satisfying an algebraic equation; points with two properties (such as lying on two different circles) translate into sets of coordinates satisfying two algebraic equations; and so on." *The Universal Encyclopedia of Mathematics* (New York: Simon & Schuster, 1964), p. 23.

2. Morris Kline, *Mathematics and Modern Culture* (Oxford: Oxford University Press, 1953), p. 186.

3. Ibid., p. 187.

CHAPTER 14: ISAAC NEWTON AND THE SEARCH FOR UNIVERSAL LAWS

1. J. D. Bernal, *Science in History* (Cambridge, Mass.: MIT Press, 1969), vol. 2, p. 462.

2. Henry Cavendish is generally credited with calculating the value of the constant of universal gravitation, which was a very small quantity expressed at 6.67×10^{-8}.

3. Newton was also a skilled craftsman as shown by his invention of the refracting telescope, a device which was regarded by his colleagues at the Royal Society as a priceless contribution to astronomy, rivaling the importance of his theoretical work in celestial mechanics.

4. John Herman Randall Jr., *The Making of the Modern Mind* (New York: Columbia University Press, 1926), p. 257.

5. Ibid., pp. 258–59.

6. See also Bernal, *Science in History*, vol. 2, pp. 481–83.

7. Richard P. Feynman, "Basic Physics," in *The Feynman Lectures on Physics* (Reading, Mass: Addison-Wesley, 1963), vol. 2, p. 2–1.

8. Max Born, "On the Meaning of Physical Theories," *Physics in My Generation* (New York: Springer-Verlag, 1969), p. 13.

9. Ibid.

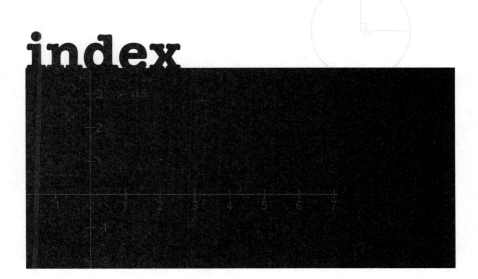

index